A
DESCRIPTION
OF THE ADMIRABLE TABLE OF
LOGARITHMES

JOHN NAPIER
BARON OF MERCHISTON
1550 – 1617

John Napier

A
Description
of the Admirable
Table of Logarithmes

A FACSIMILE OF THE
FIRST ENGLISH EDITION
TRANSLATED FROM THE LATIN
BY
EDWARD WRIGHT
PRINTED AT LONDON BY
NICHOLAS OKES
1616

RENASCENT BOOKS

First published in Latin
MIRIFICI LOGARITHMORUM CANONIS DESCRIPTIO
IOANNE NEPERO
Printed at Edinburgh by Andreæ Hart (1614)

English Translation by Edward Wright
A DESCRIPTION OF THE ADMIRABLE TABLE OF LOGARITHMES
IOHN NEPAIR
Printed at London by Nicholas Oakes (1616)
Printed at London for Simon Waterson (1618)

This facsimile edition published by
TGR Renascent Books
27 Springdale Court
Mickleover, Derby DE3 9SW
United Kingdom
2010

Paperback edition first published 2013

ISBN 978-1-4826183-1-0
www.renascentbooks.co.uk

Origination, layout and typesetting by
Gordon Roberts
Transcript and proofing by
Elizabeth Roberts

Printed and bound by
CreateSpace, Charleston, South Carolina, U.S.A.

MERCHISTON CASTLE
THE NAPIER ANCESTRAL HOME
Birthplace of Logarithms

INTRODUCTION

John Napier was born in 1550 into a wealthy and privileged family. The Napiers were people of influence in Scotland long before his birth and they were classed as belonging to the minor Scottish nobility. The family name was spelt variously at different times over a period of several hundred years as Napair, Napeir, Nepair, Nepeir, Neper, Napare, Napar and Naipper. John Napier himself used the old spelling for 'John' and commonly gave his name as 'Jhone Neper'. Ironically, the only variant of his surname he would not have seen in his lifetime was our modern spelling 'Napier'.

He was educated at St Andrew's University and later in France, possibly at the University of Paris, although this is not certain. It was whilst an undergraduate at St Andrew's that he developed his passionate and lifelong interest in theology. He was a fervent, almost fanatical, Protestant and took an active part in the religious controversies of the times. In 1593 he published *A Plaine Discovery of the Whole Revelation of Saint John*, a text which, according to the preface, he penned for the express purpose of '...preventing the apparent danger of Papistry arising within this Island.' The book considerably enhanced his reputation in Scotland and there were five editions before 1645. As early as 1607 it was translated into Dutch (three editions) and French (nine editions), a German translation (four editions) followed. As a result he acquired great prestige in Europe as a theologian. For the rest of his life Napier considered this his finest work. However, to later generations the book seemed to be a strange mixture of scholarship, bigotry and humility. It is now unread and largely forgotten.

For John Napier, the study of mathematics was only ever a hobby and it always took second place to his theological studies, running his estates and dealing with the complicated accounts associated with the management of his lands. But it was whilst engaged in the latter task that his mind began to think that there must be an easier way to do the numerous long divisions and long multiplications of the sums involved. For at least twenty years he

Introduction

plodded away, working out seemingly endless calculations in his search to eliminate the 'many slippery errors' associated with 'the multiplications, divisions, square and cubical extractions of great numbers.' In the end his studies of imaginary roots led him to invent Logarithms, a development which effectively freed the world from a logjam of calculation. Posterity would judge this to be his finest work.

Having evolved a method of simplifying tedious and difficult calculations by using sets of prepared tables, Napier decided to publish in 1614. His treatise was written in Latin, the scholarly language of his day, with the title *Mirifici Logarithmorum Canonis Descriptio* (ever after known to scholars as the *Descriptio*). Seeking a name for his discovery he turned to Greek, coining the word Logarithm from *logos* (Greek for ratio or reckoning) and *arithmos* (Greek for number). For all those people required to perform difficult and error prone calculations in the course of their working lives, such as architects, merchants, bankers and above all, astronomers, life was changed swiftly and completely. When in 1601 Johannes Kepler, the imperial mathematician at Prague, calculated the orbit of Mars – without the benefit of logarithms – it took him over four years to do it. It is not surprising therefore that he was one of the first to see the enormous importance of Napierian logarithms. However, when he first encountered them in Benjamin Ursinus' 1617 publication *Cursus Mathematicus Practicus*, after a cursory reading he unwisely dismissed them, saying they were only the work of 'Scotus Baro cujus nomen mihi excidit' (a Scottish baron whose name escapes me). But, being forced by necessity to look at logarithms again when working on his *Ephemeris*, his attitude changed from indifference to one of great enthusiasm. So pleased was he that he dedicated the *Ephemeris* to Napier, in which the unknown baron became the 'illustrious baron'. Soon, logarithms were perceived as a very useful working tool in virtually all fields requiring calculations. An example of their widespread use occurred in 1683, when the Scottish gardener John Reid published a book in which he explained how to design, calculate and layout formal gardens using logarithms.

Introduction

John Napier died on 4th April 1617 at the age of 67. He was survived by his wife Agnes, who had borne him twelve children, six of whom were still minors at his death. He died at Merchiston Castle and was buried in an aisle of the parish church of St Cuthbert's in Edinburgh. The church has been much rebuilt in the intervening centuries and today there is no grave or vault to mark his burial place. However there is a memorial tablet in the church, whose design mirrors the title page of the *Mirifici Logarithmorum*, his everlasting and immortal claim to fame. Unaware of his death, Kepler wrote to him on 28th July 1619, praising him and awarding his logarithms the highest accolades he could think of. Two hundred years later, the French mathematician and astronomer Pierre Simon Laplace said that logarithms, '...by shortening the labours, doubled the life of the astronomer.' At a congress held in Edinburgh to celebrate the 300th anniversary of the publication of logarithms, it was remarked by Lord Moulton that 'No previous work had led up to it, nothing had foreshadowed it or heralded its arrival. It stands isolated, breaking upon human thought abruptly, without borrowing from the works of other intellects or following known lines of mathematical thought.'

NAPIERIAN LOGARITHMS

In the *Descriptio*, Napier does not reveal how he calculated his logarithms, saying that he wanted to show their use rather than how he had arrived at them. However, he wrote a later work, titled *Mirifici Logarithmorum Canonis Constructio* (known simply as the *Constructio*), which was published posthumously in 1619 and in which he did explain the calculations necessary to compile tables of logarithms. Napier's logs are somewhat different from the logarithms we use today. They do not really have a base, being instead defined in terms of sines. This contrasts with modern usage of common logs to base 10 or natural logs to base e which nowadays are found on most electronic calculators (usually on keys marked 'log' and 'ln' respectively). But Napier's logarithms do involve a constant 10^7, which arose from his method of construction. He did not think of his logs in any sort of algebraic way, as we do today, since algebra

Introduction

was not sufficiently advanced in his time to make such an approach viable. Instead, he proceeded by an analogy involving dynamics.

He imagined two lines. One line, say AB, was of fixed length. The second, say $A'X$, was of infinite length. Points C and C', on AB and $A'X$ respectively, begin moving to the right simultaneously with the same initial velocity, starting at A and A'. C' continues to move with uniform velocity but C moves with a velocity that is equal to the distance CB. Napier defined $A'C'$ ($= y$) as the logarithm of BC ($= x$), that is, y = Nap. log x. His choice of 10^7 for the fixed length of AB was based on the fact that the best tables of sines available to him were given to seven decimal places, and he thought of the argument x in the form $10^2.\sin X$. A major difficulty with Napierian logarithms is that Nap. log 1 does not equal 0, which makes them much less convenient for calculations than the logs we use today.

EDWARD WRIGHT

The East India Company was not slow to appreciate Napier's discovery of logarithms. Although navigators on the ships now making great voyages had the quadrant rather than the astrolabe used a hundred years earlier by Columbus, like him they still had to carry out enormous, cumbersome and error-prone calculations to fix their position at sea. This was in a period when the multiplication and division of large numbers, as well as the taking of square roots, was considered to be extremely difficult. The company quickly realised the potential benefits which logarithms could bring to this science. They saw that logarithms could replace these difficult mathematical operations with the much simpler processes of addition, subtraction and division by two. In 1614, the very year that Napier's *Descriptio* appeared, Edward Wright was appointed as hydrographer to the company. Wright was a Cambridge educated mathematician, who had spent time at sea and had a reputation as a competent navigator. It was natural therefore, that the company should turn to him for a speedy English translation of Napier's seminal text, for (in Wright's words) '...very great use for Mariners; but also to help the want of those that could not understand it in Latin.' He completed this work early in 1615 and a copy was sent to

Introduction

Scotland for Napier's approval. Napier confirmed that he found it '...to be most exact and precisely conformable to my mind and the original.' Sadly, Wright died in late November, 1615, shortly after the manuscript was returned from Scotland. It was left to his son Samuel to see the book through publication the following year. It appeared with the title *A Description of the Admirable Table of Logarithms*, printed at London by Nicholas Okes, and it is a facsimile of this historic text which you now hold in your hands. A second English edition appeared in 1618, printed for Simon Waterson, again at London.

HENRY BRIGGS

The news of John Napier's discovery of logarithms was of such momentous importance that it quickly spread among the mathematical community. One man in particular, Henry Briggs, Professor of Mathematics at Gresham College in London, was immediately interested and impressed. Writing to Archbishop Usher in March 1615, he said, 'Napier, Lord of Merchiston, hath set my head and hands at work with his new and admirable logarithms. I hope to see him this summer if it please God, for I never saw a book which pleased me better, and made me more wonder.' Briggs meant what he said, and the following year he set forth on the arduous four day journey from London to Edinburgh. What happened on his arrival was later related to Elias Ashmole by William Lilly, who had the story from John Mair who was present at the meeting of the two men.

> When Merchiston first published his logarithms, Mr Briggs, then reader of the Astronomy (*sic*) lectures at Gresham College in London, was so surprised with admiration of them, that he would have no quietness in himself, until he had seen that noble person whose only invention they were. He acquaints John Marr, who went into Scotland before Mr Briggs, purposely to be there when these two so learned persons should meet; Mr Briggs appoints a certain day when to meet at Edinburgh, but failing thereof, Merchiston was fearful he could not come. It happened one day as John Marr and the Lord Napier were speaking of Mr Briggs: 'Ah

Introduction

John,' saith Merchiston, 'Mr Briggs will not now come.'

At that very instant one knocks at the gate; John Marr hasted down and it proved to be Mr Briggs to his great contentment. He brings Mr Briggs up into My Lord's chamber, where almost one quarter of an hour was spent, each beholding the other with admiration before one word was spoken.

At last Mr Briggs began: 'My Lord, I have undertaken this long journey purposely to see your person, and to know by what engine of wit or ingenuity you came first to think of this most excellent help unto Astronomy, viz the logarithms; but My Lord, being by you found out, I wonder nobody else found it out before, when now being known as it appears so easy.' He was nobly entertained by the Lord Napier, and every summer after that during the Laird's being alive, this venerable man Mr Briggs went purposely to Scotland to visit him.

Briggs had written to Napier before their meeting, suggesting that logs should be (in our terminology) to base 10. On this first visit he stayed for a month, and Napier agreed that new tables should be constructed with base 10 and with $\log 1 = 0$. However, he said, '...he could not, on account of ill health and for other weighty reasons undertake the construction of new tables.' Henry Briggs therefore took upon himself the tedious burden of calculating and preparing the tables, which extended to the 14th place of decimals. He completed this work in 1624. During this time he also collaborated with Napier's son Robert (the only one of the twelve children who seems to have inherited an interest in his father's mathematics), in the posthumous publication of the *Constructio*.

ABOUT THIS BOOK

Spellings

Dictionaries and standard spellings did not exist when *A Description of the Admirable Table of Logarithms* was first printed, and many words are not even spelt phonetically. Therefore many spellings in the book appear peculiar to the modern reader. Familiar words often look strange simply because, unlike modern spellings, they end with the silent letter e and the last consonant might or might not be

Introduction

doubled, hence *mane* or *manne* (man), and *rune* or *runne* (run). Note that the word logarithm is usually spelt with a final e – logarithme. The letter y is often used in place of i, for example *fynde* (find) or *fyrste* (first). Early printing conventions were to use the terminal letter s at the end of words, as today, but the long form everywhere else, for example *poſſeſs* (possess). The letters u and v were not considered to be two distinct letters, but different forms of the same letter. Typographically, v was often used at the start of words and u elsewhere, hence *vnmoued* (unmoved) or *vnloued* (unloved). But conversely, the letter v was often used where today we would expect the letter u, as in, for example, *thervnto* (thereunto). Neither were the letters i and j considered distinct, so that the word 'judge' for example would be spelt *iudge*. In this context, note the spelling of John (Napier's first name on the title page) as *Iohn*.

Diacritical Marks

Diacritical marks have been used to abbreviate printed words ever since early English printers adopted the same conventions that Gutenberg did for Latin texts. Diacritical marks are used on many pages of *A Description of the Admirable Table of Logarithms* to indicate the omission of the consonant m or n where this follows a vowel. The missing letter is indicated by placing the mark (a bar) over the vowel. Particular examples to be found in these pages are *secāt* (secant), *momēt* (moment), *thē* (them) and *educatiō* (education). There are numerous other instances and any words abbreviated with diacritical marks should be read, of course, with their full pronunciation.

Faults

Errata for *A Description of the Admirable Table of Logarithms* are to be found on the last page of the treatise. However, the attentive reader will find more faults and errors than these in its pages and, of course, no attempt has been made to indicate or remedy them in this facsimile reprint. Examples are the occasional out-of-sequence chapter numbering in the page running heads, and also mistakes in printing the correct text for some running heads, particularly in the Prefaces. Sometimes spelling mistakes occur despite the relaxed

Introduction

spelling conventions of the time, although these should more properly be regarded as typographical errors rather than mistakes in spelling as we understand the term today. Just two examples are 'prodnced' (produced, on p. 17), where the letter u has obviously been printed upside down, and 'Arithmeticaall' (Arithmetical, on page 23) which is obviously wrong. A glaring error on the title page is the second occurrence of the word 'OF', which is spelt 'OE'. There are many other examples. Some errors in the printing of the logarithm tables are listed on the page which precedes the first book (that is, before page 1). Such errors plagued printers of tables for centuries, for as long as human skill (or lack of it) was a major factor in typesetting. The tables of logarithms printed in *A Description of the Admirable Table of Logarithms* are no exception and there are many more than those given on the aforementioned page. Typically, errors occur when numbers are juxtaposed so that a series of logarithms beginning, say, with 93... are printed as 39... Again, 6 and 9 are often printed upside down, rendering a number totally different to the one intended. When these and similar typographical and/or mathematical errors are encountered in this book, remember that they occur in the original printing and are faithfully reproduced in this facsimile reprint.

Plus and Minus Signs

The mathematical symbols for plus + and minus −, used by the 17th century printers of this book, are larger than those commonly used today and are not always of consistent size. Unfortunately, the printers also used a rule, a thick black line to guide the eye across an intervening space between words and/or numbers, which is easily confused with the printed minus sign. Therefore, where numbers are concerned it is not always clear whether a rule or a minus sign is intended. However, the matter is usually resolvable by diligent examination of the mathematics and of the context in which the line is used.

The Dedication

Napier dedicates his book to 'the most noble and hopeful Prince Charles, only son of the high and mighty James....' This prince, of

Introduction

the house of Stuart and fourteen years old when the *Descriptio* was published, was crowned King of England in March 1625. Napier did not live to see his coronation, but at the end of the dedication he calls the young prince '...the hope of our future tranquilitie.' He could have had no premonition that this young man would in adulthood plunge his people into two civil wars, hasten the abolition of the monarchy, be beheaded for treason and, for the first and only time in England's history, precipitate a republic called the Commonwealth of England under Lord Protector Oliver Cromwell.

The Instrumental Table

While Edward Wright was working on the translation of the *Descriptio*, it must have occurred to him that the table of logarithms there presented could not contain every number that might be required by a navigator at sea. Aware as he was of the easy familiarity with which seamen used compass and rule to plot their courses on charts, he devised a type of chart which he called 'An Instrumental Table for the Finding of the Part Proportional', whereby similar techniques could be employed to supplement the logarithmic tables. According to Wright's son Samuel, Napier saw and approved of the Instrumental Table. But Wright's intention to pen a description and explanation of its use was frustrated by his death. However, the indefatigable Henry Briggs undertook the task, and the Instrumental Table with his explanatory text was printed as an appendix to *A Description of the Admirable Table of Logarithmes*. This appendix, of course, did not appear in the Latin *Descriptio* of 1614.

SOURCES

Readers wanting to know more about John Napier and logarithms should consult the following:

For easily accessible biographies of Napier and Henry Briggs visit the MacTutor History of Mathematics, [online]
http://www-history.mcs.st-and.ac.uk/Mathematicians/Napier.html
http://www-history.mcs.st-and.ac.uk/Mathematicians/Briggs.html

Written sources are:

Molland, George, 'Napier, John, of Merchiston (1550–1617)', *Oxford Dictionary of National Biography*, Oxford University Press. (Sept 2004).

Introduction

Kaunzner, Wolfgang, 'Briggs, Henry (*bap. 1561, d. 1631)*', *Oxford Dictionary of National Biography*, Oxford University Press. (Sept 2004).

Apt, A. J. 'Wright, Edward (*bap. 1561, d. 1615)*', *Oxford Dictionary of National Biography*, Oxford University Press. (Sept 2004).

Carslaw, H. S. 'The discovery of logarithms by Napier of Merchistoun', *Journal of Proceedings of the Royal Society of New South Wales*, 48 (1914), 42–72.

Gibson, G.A. 'Napier and the Invention of Logarithms' in *Proceedings of the Royal Philosophical Society of Glasgow*, (1914).

Gladstone-Millar, Lynne, *John Napier – Logarithm John*, National Museums of Scotland Publishing (2003).

Hawkins, W. F. 'The mathematical work of John Napier, 1550–1617', *Bulletin of the Australian Mathematical Society*, 26 (1982), 455–68.

Knott, C.G. Ed., *Napier Tercentenary Memorial Volume*, Royal Society of Edinburgh (1915).

Napier A.S. *A History of the Napier's of Merchiston*, (1923)

Napier, P. *A Difficult Country – The Napier's in Scotland*, Michael Joseph, London (1972).

Shennan, F. *Flesh and Bones – The Life, Passions and Legacies of John Napier*, Napier Polytechnic of Edinburgh (1989).

HERE BEGINS
A
DESCRIPTION
OF THE ADMIRABLE TABLE OF
LOGARITHMES

A
DESCRIPTION
OF THE ADMIRABLE
TABLE OE LOGA-
RITHMES :

WITH

A DECLARATION OF
THE MOST PLENTIFVL, EASY,
and fpeedy vfe thereof in both kindes
of Trigonometrie, as alfo in all
Mathematicall calculations.

INVENTED AND PVBLI-
SHED IN LATIN BY THAT
Honorable L. IOHN NEPAIR, Ba-
ron of *Marchiston,* and tranflated into
Englifh by the late learned and
famous Mathematician
Edward Wright.

*With an Addition of an Inftrumentall Table
to finde the part proportionall, inuented by
the Tranflator, and defcribed in the end
of the Booke by* HENRY BRIGS
Geometry-reader at Grefham-
houfe in London.

All perufed and approued by the Author, & publifhed fince the death of the Tranflator.

LONDON,
Printed by NICHOLAS OKES.
1616.

TO THE RIGHT
HONOVRABLE AND
RIGHT WORSHIPFVLL
COMPANY OF MERCHANTS
of London trading to the Eaft-
Indies, SAMVEL WRIGHT
*wifheth all profperitie in this
life, and happineffe in the
life to come.*

Our fauours towards
my deceafed Father,
and your imployment
of him in bufineffe of
this nature, but chiefe-
ly your continuall im-
ployment of fo many Mariners in fo
many goodly and coftly fhips, in long
and dangerous voyages, for whofe vfe
(though many other wayes profitable)
this little booke is chiefly behoouefull :
may chalenge an intereft in thefe his
labours. This *Book* is noble by birth, as
being defcended from a Noble Parent,
& not ignoble by educatiõ, hauing learned to fpeake Englifh of my late Fa-
A 2   ther,

ther, a man in the iudgment of the learned, and experience of the common
fort, famous for knowledge and practife in the Mathematickes : whofe
care thereof was fo great, to fend it a-
broad with the true refemblance of his
worthy father, and fufficient knowledge of the Englifh tongue to inftruct
our Countrey-men, that hee procured
the Authors perufall of it : who after
great paines taken therein, gaue approbation to it, both in fubftance and
forme, as now I prefent it vnto you. I
am the bolder thus to do, in regard it is
not vnknowne to many men, that my
faid father fpent a great part of his time
in ftudy of the Art of Nauigation, and
had gathered much vnderftanding by
is owne practife in fome voyages to
fea with the right Honourable the
Earle of *Cumberland* deceafed : where-
upon he publifhed a painful worke dif-
couering errours committed by Mariners in that Art, with corrections and
ready wayes for reformation therof. So
that I thinke it is out of doubt, that his
iudgement therein was great. And feeing hee not onely gaue much commen-
dation of this worke (and often in my
hearing) as of very great vfe for Mari-

ners :

ners : but alſo to help the want of thoſe that could not vnderſtand it in Latine, tranſlated the ſame into Engliſh, and added thereto an inſtrumentall Table to finde the part proportional, whereof alſo the noble Author approued well. I doubt not but it is apparant enough that he eſteemed of it, and intended to haue recommended it as a booke of more then ordinary worth, eſpecially to Sea-men. But ſhortly after he had it returned out of *Scotland,* it pleaſed God to call him away afore he could publiſh it, or but write a deſcription of the ſaid inſtrumentall Table which he had deuiſed, therefore hee left the publiſhing of it to me, as an inheritance, and the ſaid deſcription to his learned and kind friend Mr. *Henry Brigges,* who hath performed it accordingly. All which I humbly preſent vnto you, hoping you ſhall receaue as much profite by the vſe of it, as there hath been learning, care, and paines beſtowed in the penning and fitting it thus to your hands.

A3

TO THE MOST
NOBLE AND HOPE-
FVLL PRINCE,

CHARLES:

ONELY SONNE OF

the high and mightie IAMES by
the grace of God, King of great Brittaine, France, and Ireland : Prince
of Wales : Duke of Yorke and
Rothefay : Great Steward of
Scotland : and Lord of
the Iflands.

MOST NOBLE PRINCE,

EEING there is neither
study, nor any kinde of lear-
ning that doth more acuate
and stirre up generous and
heroicall wits to excellent
and eminent affaires : and
contrariwise that doth more deiect and keepe
downe fottifh and dull mindes, then the Ma-
thematickes. It is no maruell that learned
and

Dedicatorie.

and magnanimous Princes *in all former ages haue taken great delight
in them, and that vnſkilfull and ſlothfull
men haue alwayes purſued them with
moſt cruell hatred, as vtter enemies to
their ignorance and ſluggiſhneſſe. Why then
may not this my new inuention (ſeeing it ab-
horreth blunt and baſe natures) ſeeke and
flye vnto your Highneſſe moſt noble diſpoſi-
tion and patronage? and eſpecially ſeeing this
new courſe of* Logarithmes *doth cleane
take away all the difficultie that heretofore
hath beene in mathematicall calculations,
(which otherwiſe might haue beene diſtaſt-
full to your worthy towardneſſe) and is ſo
fitted to helpe the weakneſſe of memory, that
by meanes thereof it is eaſie to reſolue moe
Mathematical queſtions in one houres ſpace,
then otherwiſe by that wonted and commonly receiued manner of Sines, Tangents, and
Secants, can bee done euen in a whole day.
And therefore this inuention (I hope) will
bee ſo much the more acceptable to your
Highneſſe, as it yeeldeth a more eaſie and
ſpeedy way of accompt. For what can bee
more delightfull and more excellent in any
kinde of learning then to diſpatch honourable and profound matters, exactly, readily,
and without loſſe of either time or labour. I
craue therefore (moſt gracious* Prince) *that*
you

Dedicatorie.

you would (according to your gentleneffe) accept of this gift) though fmall, and farre beneath the height of your deferuings, and worth) as a pledge and token of my humble feruice : which if I vnderftand you doe, you fhall (euen in this regard onely) encourage me that am now almoft fpent with fickneffe, fhortly to attempt other mattters, perhaps greater then thefe, and more worthy fo great a Prince. In the meane while, the fupreame King of Kings, and Lord of Lords long defend and preferue to vs the great lights of great Brittaine, your renowned parents, and your felfe the noble branch of fo noble a ftemme, and the hope of our future tranquilitie : to him be giuen all honour and glory.

Your Highneffe moft
deuoted Seruant,

IOHN NEPAIR.

The Authors Preface to
the Admirable Table of Logarithmes,

EEING there is nothing (right well beloued Students in the Mathematickes) that is so troublesome to Mathematicall practise, nor that doth more moleft and hinder Calculators, then the Multiplications, Diuisions, square and cubical Extractions of great numbers, which besides the tedious expence of time, are for the most part subiect to many slippery errors. I began therefore to consider in my minde, by what certaine and ready Art I might remoue those hindrances. And hauing thought vpon many things to this purpose, I found at length some excellent briefe rules to be treated of (perhaps) hereafter. But amongst all, none more profitable then this, which together with the hard and tedious Multiplications, Diuisions, and Extractions of rootes,

A 5   doth

The Authors Preface.

doth alſo caſt away from the worke it ſelfe, euen the very numbers themſelues that are to be multiplied, diuided, and reſolued into rootes, and putteth other numbers in their place, which performe as much as they can do, onely by Addition and Subtraction, Diuiſion by two, or Diuiſion by three : which ſecret inuention, being (as all other good things are) ſo much the better as it ſhall be the more common, I thought good heretofore to ſet forth in Latine for the publique vſe of Mathematicians. But now ſome of our Countreymen in this Iſland well affected to theſe ſtudies, and the more publique good, procured a moſt learned Mathematician to tranſlate the ſame into our vulgar Engliſh tongue, who after he had finiſhed it, ſent the Coppy of it to me, to bee ſeene and conſidered on by my ſelfe. I hauing moſt willingly and gladly done the ſame, finde it to bee moſt exact and preciſely conformable to my minde and the originall. Therefore it may pleaſe you who are inclined to theſe ſtudies, to receiue it from me and the Tranſlator, with as much good will as we recommend it vnto you. Fare yee well.

THE PREFACE
TO THE READER

By *Henry Brigges.*

ENTLE Reader, feeing I
haue publickly taught the
meaning & vfe of this booke
at *Grefham* houfe, and haue
had fome charge about this Impreffion
committed vnto me, both by the Honourable Authour the L. of *Marchi-
ſton,* and by my very good frend M^r. *Edward Wright* the Tranſlator. And feeing
the one who hath moſt right, and is
beſt able to commend it, is fo farre ab-
fent, and the other hath made a moſt
happy change of this place and life for
a better : thou maiſt happily expect that
I ſhould write fomewhat that may giue
fome taſte of the excellent vfe of it to
thofe who by reafon of the diſtance of
place, or other occaſions, cannot come
to heare me. In a word therefore I will
bee bold to fet downe mine opinion,
writing nothing but that which I hope
I ſhall alwayes be able and willing to
make good and maintaine. There hath
beene

The Authors Preface.

been for many former ages euen vnto
this prefent, a very great deale of time
and expences beftowed by moft indu-
ftrious, learned, and worthy men about
the doctrine of Triangles, and the making of the Tables of *Sines*, *Tangents*,
and *Secants*, that by the helpe of them
we may attaine to the knowledge and
vfe of the Mathematickes, and efpecially of Aftronomie and Nauigation, as
namely by *Hipparchus*, *Ptolomey*, *Theon*,
Regiomontanus, *Copernicus*, *Reinoldus*, *Fin-
kius*, *Lansbergius*, *Clauius*, *Adr. Romanus*,
Ioach. Rheticus, *Valent Otha*, and *Pitifcus*.
All thefe, and diuers others, to their
exceeding praife, and the great eafe &
contentment of all fuch as fet themfelues to the ferious ftudie of the Mathematickes, haue laboured much, and
fome of them beftowed very great coft,
both of their owne eftate, & alfo from
the liberall contribution of fundry
great Princes vpon the maintenance of
diuers men, who for many yeares together haue wholly employed themfelues
to calculate thefe Tables. Yet notwith-
ftanding this little Table of Logarithmes being firft begun, and finifhed by
the charge and paines of the honourable Authour alone, may for exactneffe
and

M^r. Briggs Preface.

and certainty compare with all thofe
Tables, and for eafe and expedition go
very farre beyond them, for all Trigo-
nometricall operations, efpecially
Sphæricall, and for the making of the
Tables of *Profthaphærefes* for the Planets. Which confiderations may iuftly
warrant the Title of *The Admirable Table of Logarithmes.* But befides all this,
there is an other very excellent and admirable vfe of this Table, which is not
at all furthered by the other Tables
formerly mentioned, nor can (for any
thing I know) be any other way performed, but with very great paines and
loffe of much time : and that is in numbers continually proportionall, hauing
any two numbers giuen with their diftance, or with the number of meane
proportionals betwixt them, at one o-
peration to find any one of thofe meane
proportionals, or any one of the numbers, without the giuen numbers at a-
ny diftance affigned. And becaufe thefe
things may to fome feeme obfcure, giue
me leaue to explaine thē by an *example.*
Let the two giuen numbers be 1. and
3000. and let there be fuppofed foure
meane proportionals betwixt them. If
of thefe foure I defire that which is neereft

M. *Briggs Preface*.

reſt to the leſſe extreame, that meane,
(becauſe here the leſſe extreme is an v-
nity) is called the ſurſolide root of the
other extreame, to wit, of 3000. and
that, or any other root may farre more
eaſily be had by theſe *Logarithmes* then
by any rule or other way. But the finding of any root is but one particular
meane proportionall, to wit, the
next meane to the vnity : and this way
is generall, giuing as eaſily the third or
fourth meane as the firſt. And not onely where the one extreame is an vnity,
but betwixt any two numbers aſſigned.
For example, if the giuen extremes bee
19 and 738. and there ſhall be betwixt
theſe two, ten meane proportionals : by
this Table we may finde the 7ᵗʰ or 8ᵗʰ,
or any other aſſigned, from the leſſe or
from the greater number : or if they be
continued further either diminiſhing
vnder 19. or increaſing aboue 738. we
may finde any of them for any diſtance
aſſigned, as the fifth or ſixth in the ſame
proportiõ aboue 738, or vnder 19. And
thus hauing two extremes giuen, and
the number of meane proportionals betwixt them, we may finde any, for any
aſſigned diſtance within or without. In
like ſort, hauing a proportion aſſigned

in

M. *Briggs* *Preface*.

in numbers, and a third number giuen,
we may from that third number find an
extreme : betwixt which and that third
number fhall bee any fet number of
meane proportionals, keeping the proportion affigned one from another. For
example,if 73 ꝑ be yearely fo to be in-
creafed, as that $\frac{1}{16}$ be ftill to bee added
vnto the former yeares fumme, and I
would know what is the whole at the
end of feuen years : here the proportion affigned is 16 to 17 the third number, or the beginning of the progreffi-
on is 73, the meane proportionals are
fixe : I would know the other extreme,
to wit, the feuenth from the beginning,
and by this Table of *Logarithmes* I find
it to bee 111 ℔, 11 ẞ, 9 ◌ $\frac{696}{1000}$, which
perhaps by curious fearch, after the la-
borious ordinarie way, will bee found
too great by $\frac{1}{11}$ of a peny, or therea-
bouts. And thus we fee the admirable
vfe of thefe *Logarithmes*, not onely in
the doctrine of Triangles (which I account to bee farre the moft excellent
part, and which may by other Tables be
performed as exactly, but nothing fo
fpeedily, or with the like eafe) but alfo
in all our common accounts of ordinary
proportionall numbers : wherein wee
may

M͏ᷓ. Briggs Preface.

may not expect the fame exactneſſe
which we may attaine vnto by rule, and
by long tedious practiſe, becauſe this
table is but finall, and the numbers neuer exceed the eighth place; but wee
may ſafely truſt to it to performe all
things without ſenſible errour : or to
performe truly, ſo farre as can be expreſſed in 7 figures. And if it ſhall pleaſe
God (who beſides his other mercies
hath granted this honour vnto the Authour, to begin and thus farre to accom-
pliſh this admirable worke) further to
grant vnto him life and competent
ſtrength, I doubt not we ſhall haue the
worke ſo enlarged and perfected, that
we may vſe it, both with greater eaſe,
& with exactneſſe vnto the 10ᵗʰ place.
And thus commending theſe things to
thy conſideration, and vs and all our
honeſt ſtudies to the Lords bleſſing, I
euer reſt a louer of all them that loue
the Mathematickes,

H. Brigges.

In praiſe of the neuer-
too-much praiſed Worke and
Authour the L. of
Marchiſton.

*W*Hat, lik̲e *our bodies,* ſoules *rare* excellence,
  Our bodies bound, yet haue thereof no ſght.
(*Enwomb'd with* clouds *of* Myſtery *from ſence*)
Is here (*well* borne, *and* ſhap't) *produc'd to light;*
This skill, *ſince firſt men* k̲new, *ſtill lay* ynknowne.
As if ſome meere Impoſſibilitie
Had ſtood twixt It *and how it might be ſhowne.*
But now it look̲es lik̲e ſelfe-Facilitie !
How happy *and* acute *were his* Wits *eyes,*
That for the Mathematicks *found this* Key,
To ope the lockes *of all their* Miſteries,
That from all eyes ſo long concealed *lay.*
It was at hand, *and yet it was* vnſeene:
Inuiſible, *and yet was* cleere *to* wit
As it could wiſh,or as it could haue beene
In Art *or* Nature; *yet* Art *miſt of* It.
From whence a queſtion *may ariſe* (*perchance*)
Whether, or no, This *do extenuate*
The Authors *merit?* N̲o̲, *it doth aduance*
His praiſe the more, the leſſe he toil'd for that.
For who with eaſe hath done what none *ere could*
Is moſt lik̲e God *in* workes *of* rareſt ſkill,
This argues He *can do what ere he would*
In Art *with* eaſe, *if he had but a* Will. (*then,*
∗Wright (*ſhip-wright? no; ſhip-right, or righter*   M. *Wrights*
when wrong ſhe goes) *lo this, with eaſe, will make*   Tract of
Thy Rules *to mak̲e the* ſhip *run rightly,when*   Nauigati-
                                                She on.

She thwarts the Maine *for* Praiſe *or* profits *ſake.*
If after-times, *that ſtill ſhall bleſſe his* name,
Shall ſeeke more eaſe *than, in his* eaſineſſe,
To worke by Figures, *he muſt make* Art *lame*
(So leſſe deſir'd) with Eaſes *great exceſſe!*
For his Rules *are ſo* firme *and* facill *too,*
As makes Art *laugh their* quick-diſpatch *to waigh*
With Tangents *and with* Secants *much-a-do,*
And Enuy *with that* eaſe *to pine away.* (*ſure*
O that great Lords *no worſe would vſe their lea-*
In ſeuerall kinds, then (kindly) were they Great:
*But they make ſmall thēſelues w*ᵗʰ *too-great pleaſure:*
*So, great-*Lords *th'are not, nor their* Counterfet.

Scotland, *two* Miracles *of* Men, *this* Age
In thee affoords the world, *to future yeares:*
Bucanan. *The* Tutor *of our* Rulers Pupillage,
And this rare Lord, *a* Lead-ſtarre *to his* Peeres.
The ground of whoſe iuſt praiſes is ſo ſure, (wing:
That it will beare more Fame *then* Fames *right*
Birth, Grace, *and* Art, *and all ſurpaſſing pure,*
Makes *him more* good *then* great, *although a* king.
Then great good Lord, *liue euer in my* Lines,
By thy iuſt lauds *that ſhall then (dead) reuiue,*
Vnill *the* Sunne *forſake the heauenly* Signes,
And in the Signes *of thy* worth *euer liue.* (thee,
To light the world *through* them, & *them* through
And me through both, to Fame, & *that through me!*

By the vnfained louer and
admirer of his Art *and*
matchleſſe virtue,

Iohn Dauies *of* Hereford.

In the iuſt praiſe of this
Booke, Authour, and
Tranſlator.

A Rts, *in theſelues, haue ſuch diuine* Perfection,
As Human reaſon *cannot alwaies ſee;*
Yet God *all good, to man giues ſuch direction*
As hidden things *ſometimes* diſcouered *bee:*
    What many men *and* ages *could not finde,*
    Is, at the laſt, by ſome one *brought to mind.*

This noble Author *firſt due* honour *gaue*   By his
To him from whom true honours doe proceed, *   worke on
Who now to him doth graciouſly vouchſaue   the *Reuela-*
Beſides his Stile, *much* honour *for his* meed,   *tion,* first
    By bringing Him *thus clearly to reueale*   printed in
    Such profit *both to* Church *& Comon*-weale   *An* 1593.
                                                     And a-
This little Booke *(to let the other paſſe)*   gaine in
As Title *ſhewes, is truly admirable,*   *An* 1611.
Th'inuention *rare, for* practiſe *nothing leſſe,*
Briefe, eaſie, plaine, *and paſſing* delectable.
    What earſt was hard *and* tedious *to vnfold,*
    Here how to find with eaſe, *is* plainly *told.*

The toyleſome Rules *of due* Proportion
Done here by Addition *and* Subtraction,
By Bipartition *and* Tripartition,
The Square *and* cubicke rootes *extraction:*
    And ſo, all queſtions Geometricall,
    But with moſt eaſe Triangles-ſphæricall.

The

The vſe *is great in all true* Meaſuring
Of Lands, Plots, Buildings, *and* Fortification,
So in Aſtronomie *and* Dialling,
Geography *and* Nauigation.
 In theſe *and like, yong ſtudents ſoone may gaine,*
 The ſkilfull too, may ſaue coſt, time *&* paine.

In Latine *to the* world *it firſt appear'd*

Mʳ *Wright*
detected
& corre-
cted ma-
ny errors
in the vul-
gar Naui-
gation.

Strange vnto them to whom that tongue *is ſtrange:*
But he who earſt our Nauigation *clear'd,*
From that ſtrange tongue to Engliſh *it did change.*
 That famous, learned, Errors *true correƐor,*
 England, *great* Pilot, Mariners Director.

Whoſe care thereof was ſuch, that he obtain'd
The Authors Approbation, *and withall,*
He, for the helpe of Practiſers *ordain'd*
A way to finde the part Proportionall:
 The yſe whereof too-timely death *deny'd,*
 Which famous Briggs *hath learnedly ſupply'd.*

Thus haue you here the quinteſſence *of* Art,
Fitted to hand by men of rareſt skill,
Whoſe euerlaſting praiſes *in each part*
So farre extend that here conclude *I will.*
 And ſay; For Matter, Author, *and* Tranſlator,
 Nere had theſe Arts *ſo good a* Demonſtrator.
 Pulchra hæc facilia.

Ri. Leuer.

A VIEW OF THIS BOOKE.

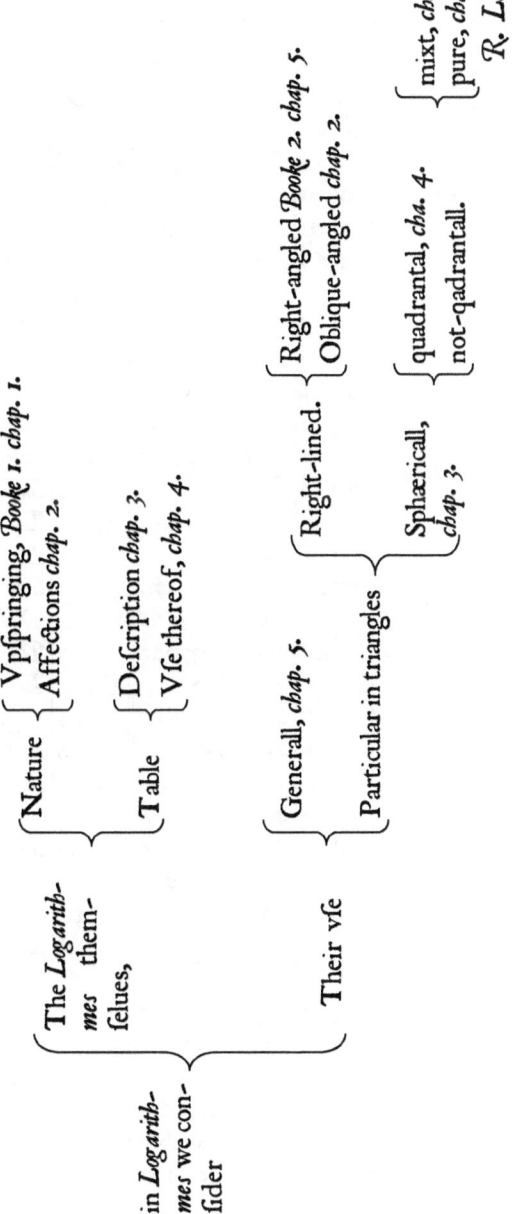

Some faults haue escaped in printing of the Table, which the practiser (if it please him to take the paines) may easily correct afore he vse the Table, after this manner, or what else he shall find.

Deg.	mi		
0	10	Diff.	5839981
0	11	Sine	3200
1	21	Sine	23560
1	60	Log.	3355282
2	1	Log.	3346986
4	46	Log.	2487733
85	0	Sine	996195
83	44	Sine	994025
	3	Sine	992652
8	33	Diff.	1894833
81	30	Sine	989015
11	18	Sine	195946
	24	Log.	1621220
78	42	Log.	19576
13	45	Sine	237686
	46	Sine	237968
	47	Sine	238251
	48	Sine	238533
	49	Log.	1432062
	50	Log.	1430880
15	5	Sine	260224
16	26	Diff.	1220955
21	6	Sine	359997
22	4	Diff.	902930
	50	Log.	946616
23	2	Log.	938366
24	58	Diff.	764429
64	2	Sine	899049
27	4	Sine	455027
	11	Sine	456839
	20	Diff.	659953
	31	Sine	462007
62	30	Sine	887011

Deg.	m		
62	18	Sine	885392
28	34	Diff.	607955
61	1	Sine	874761
60	47	Log.	136071
	60	Sine	874620
29	59	Diff.	549978
60	20	Log.	140504
	21	Log	140339
	7	Log.	142668
57	30	Sine	843391
	14	Sine	840882
33	59	Log	581692
54	5	Log.	210818
36	28	Sine	594355
53	1	Diff	283560
	7	Diff.	287193
52	15	Log.	234850
38	26	Log.	475452
	59	Log.	463474
51	0	Sine	777146
	4	Sine	777878
	19	Sine	780612
38	49	Log.	467079
	44	Log	468889
39	13	Sine	632255

I

A DESCRIPTI-
ON OF THE ADMIRABLE
TABLE OF LOGARITHMES,
WITH THE MOST PLEN-
TIFVL, EASIE, AND READY
Vſe thereof in both kindes of
Trigonometrie, as alſo in all Ma-
thematicall Accounts.

THE FIRST BOOKE.

CHAP. I.
Of the Definitions.

 LINE *is ſaid to increaſe equally,*   1. *Definiti-*
when the poynt deſcribing the ſame,   *on.*
goeth forward equall ſpaces, in
equall times, or moments.

Momèt 1	2	3	4	5	6	7	8	9	10	11	12	
A	C	D	E	F	G	H	I	K	L	M	N	O &c
B	B	B	B	B	B	B	B	B	B	B	B	

Let A be a poynt, from which a line is to be
drawne by the motion of another poynt,
which let be B.

Now in the firſt moment, let B moue from

B   A

A to C. In the ſecond moment from C to D. In the third moment from D to E, & ſo forth infinitely, deſcribing the line ACDEF, &c. The ſpaces AC, CD, DE, EF, &c. And all the reſt being equall, and deſcribed in equall moments (or times.) This line by the former definition ſhall be ſaid to increaſe equally.

A Corollary or conſequent.

Therefore by this increaſing, quantities equally differing, muſt needes be produced, in times equally differing.

As in the Figure before, B went forward from A to C in one moment, and from A to E in three moments. So in ſixe moments from A to H: and in 8 moments from A to K. And the differences of thoſe moments, one and three, and of theſe 6 and 8 are equall, that is to ſay two.

So alſo of thoſe quantities AC, and AE, and of theſe, AH, and AK, the differences CE, and HK are equall, and therefore differing equally, as before.

2. Definition.

A line is ſaid to decreaſe proportionally into a ſhorter, when the poynt deſcribing the ſame in æquall times, cutteth off parts continually of the ſame proportion to the lines from which they are cut off.

S R   Q
├─┼────────────────────────┤

```
        b    b    b    b   b  b  b b bbb
 ●──────●────●────●────●───●──●──●●●●●●────Z
 a      c    d    e    f   g  h  i k lmno
 Momèt 1     2    3    4   5  6  7 8 9 10 11 12  &c
```

For examples ſake. Let the line of the whole ſine a Z be to bee diminiſhed proportionally : let the poynt diminiſhing the ſame by his motion

CHAP.I.   *The firſt Booke.*   3

motion be *b* : and let the proportion of each part to the line from wᶜʰ it is cut off, be as Q R to Q S. Therefore in what proportion Q S is cut in R, in the ſame proportion(by the 10 of the *6* of *Euclid*) Let *a* Z be cut in *c.* and ſo let *b.* running from *a* to *c* in the firſt moment, cut off *a c* from *a* Z, the line or ſine *c* Z remaining.

And from this *c* Z let *b* proceeding in the ſecond moment, cut off the like ſegment, or part, as QR to QS : and let that bee *c d,* leauing the ſine. *d* Z. From which therefore in the third moment, let *b* in like manner, cut off the ſegment *d e,* the ſine *e* Z being left behinde. From which likewiſe in the fourth moment, by the motion of *b,* let the ſegment *e f* be cut off, leauing the ſine *f* Z. From this *f* Z in the fifth moment, let *b* in the ſame proportion cut off the ſegment *f g,* leauing the ſine *g* Z, and ſo forth infinitly. I ſay therfore out of the former definition, that here the line of the whole ſine *a* Z, doth proportionally decreaſe into the ſigne *g* Z, or into any other laſt ſine, in which *b* ſtayeth, and ſo in others.

Hence it followeth that by this decreaſe in e-   *A Corolary.*
quall moments (or times) there muſt needes alſo bee left proportionall lines of the ſame proportion.

For what continuall proportion there is before of the ſines to be diminiſhed, *a* z, *c* z, *d* z, *e* z, *f* z, *g* z, *h* z, *i* z, and *k* z, &c. and of the ſegments cut off from them, *a c, c d, d e, e f, f g, g h, h i,* and *i k* there muſt needes be alſo the ſame proportion of the ſines remaining, that is, *c* z, *d* z, *e* z, *f* z, *g* z, *h* z, *i* z, and *k* z, as may manifeſtly ap-

4   *The firſt Booke.*   CHAP.I.

peare by the 19 *Prop.* 5. and 11. *Prop.* 7, *Eu-clid.*

3 Def.

Surd quantities, or vnexplicable by number, are ſaid to be defined, or expreſſed by numbers very neere, when they are defined or expreſſed by great numbers which differ not ſo much as one vnite from the true value of the Surd quantities.

As for example. Let the ſemidiameter, or whole ſine be the rational number 10000000 the ſine of 45 degrees ſhall be the ſquare root of 50,000,000,000,000, which is ſurd, or ir-rationall and inexplicable by any number, & is included between the limits of 7071067 the leſſe, and 7071068 the greater : therfore, it differeth not an vnite from either of theſe. Therefore that ſurd ſine of 45 degrees, is ſaid to be defined and expreſſed very neere, when it is expreſſed by the whole numbers, 7071067, or 7071068, not regarding the fra-ctions. For in great numbers there ariſeth no ſenſible error, by neglecting the fragments, or parts of an vnite.

4 Def.

Equall-timed motions are thoſe which are made together, and in the ſame time.

As in the figures following, admit that B be moued from A to C, in the ſame time, wherin *b* is moued from *a* to *c* the right lines AC & *a c,* ſhall be ſayd to be deſcribed with an e-quall-timed motion.

5 Def.

Seeing that there may bee a ſlower and a ſwif-ter motion giuen then any motion, it ſhall neceſſa-rily follow, that there may be a motion giuen of e-quall ſwiftneſſe to any motion (which wee define to be neither ſwifter nor ſlower.)

6 Def.

The Logarithme *therfore of any ſine is a num-ber very neerely expreſſing the line, which increa-ſed*

Chap.2.   *The firſt Booke.*   5

ſed equally in the meane time, whiles the line of
the whole ſine decreaſed proportionally into that
ſine, both motions being equal-timed, and the beginning equally ſwift.

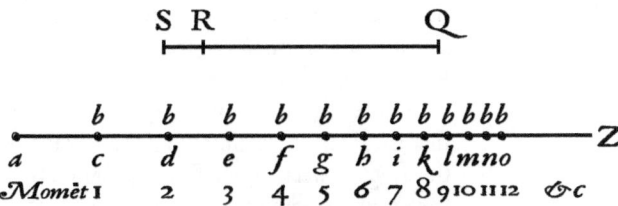

As for example. Let the 2 figures going afore
bee here repeated, and let B bee moued alwayes, and euery where with equall, or the
ſame ſwiftneſſe wherewith *b* beganne to bee
moued in the beginning, when it was in *a*.
Then in the firſt moment let B proceed from
A to C, and in the ſame time let *b* moue proportionally from *a* to *c*, the number defining or expreſſing AC ſhal be the *Logarithme*
of the line, or ſine c Z. Then in the ſecond
moment let B bee moued forward from C to
D. And in the ſame moment or time let *b* be
moued proportionally from *c* to *d*, the
number defining AD ſhall bee the *Loga-*
rithme of the ſine d Z. So in the third moment let B go forward equally from D to E,
and in the ſame moment let *b* be moued forward proportionally from *d* to *e*, the number expreſſing AE the *Logarithme* of the ſine
c Z. Alſo in the fourth moment, let B pro-
B 3   ceed

6   *The firſt Booke.*   Chap.i.

ceed to F, and *b* to *f,* the number AF ſhall
be the *Logarithme* of the ſine *f z.* And keeping the ſame order continually (according
to the former definition) the number of AG
ſhall be the *Logarithme* of the ſine *g z.* AH
the *Logarithme* of the ſine *h z.* AI the *Loga-*
rithme of the ſine *i z.* AK the *Logarithme* of
the ſine *k z,* and ſo forth infinitely,

A cōſequēt.   *Therefore the Logarithme of the whole ſine*
1000000 is nothing, or o : and conſequently the
Logarithmes of numbers greater then the whole
ſine, are leſſe then nothing.

For ſeeing it is manifeſt by the definition,
that the ſines decreaſing from the whole ſine,
the *Logarithmes* increaſe from nothing : therfore contrariwiſe the numbers which yet we
call Sines, increaſing vnto the whole ſine,
that is to 1000000 , the *Logarithmes* muſt
needs decreaſe to o or nothing : and by
conſequent the *Logarithmes* of numbers increaſing aboue the whole ſine 1000000,
which wee call *Secants,* or *Tangents,* and no
more ſines, ſhall be leſſe then nothing.

Therefore we call the Logarithmes of the ſines
Abounding, becauſe they are alwayes greater then
nothing, and ſet this marke ╋ *before them, or*
elſe none. But the Logarithmes which are leſſe
then nothing, we cal Defeĉtiue, or wanting, ſetting
this marke ── *before them.*

It was indeed left at libertie in the beginning, to attribute nothing, or o. to any ſine
or quantitie for his Logarithme : but it was
beſt to fit it to the whole ſine, that the Addition or Subtraĉtion of that Logarithme
which is moſt frequent in all Calculations,
might neuer after be any trouble to vs.

Chap.

CHAP. II.
Of the Propoſitions of Logarithmes.

HE Logarithmes of *Proporti-* *Propoſ. 1.*
onall numbers and quantities
are equally differing.
As for example. The Lo-
rithmes of the proportio-
nall ſines, namely *c z,*
which is to *e z,* as *h z* is to
k z, are reſpectiuely the numbers defining
AC, AE, AH, AK, (as is manifeſt by the
the *6* Definition.) Now AC, and AE differ
by the difference CE, and AH and AK
by the difference HK. But by the firſt definition and his Corolarie CE and HK, are
equall : therefore the *Logarithmes* of the foreſaid proportional ſines are equally differing.
And ſo in all proportionals.

For what affections and ſymtomes the *Lo-*
garithmes haue gotten in their firſt beginning
and generation, the ſame muſt they needes
retaine and keepe afterwards.

But in their beginning and generation,
they are indued with this affection, and this
law is preſcribed vnto them, that they bee e-
qually differing, when their ſines or quantities are proportionall (as it appeareth by the
definition of a *Logarithme,* and of both motions, and ſhall hereafter more fully appeare
in the making of the *Logarithmes.*) Therefore
the *Logarithmes* of proportional quantities
are equally differing.

Of the Logarithmes of three proportionals, the *Propoſ. 2.*
double of the ſecond or meane, made leſſe by the
firſt, is equall to the third.

B 4 Seeing

8   *The firſt Booke.*   CHAP.I.

Seeing that by the firſt propoſ. the difference of the *Logarithme* of the firſt and ſecond, is equall to the difference of the *Logarithmes* of the ſecond and third, that is, the
ſecond made leſſe by the firſt, is equall to the
third, leſſe by the ſecond : therefore the ſecond being added to both ſides of the equa-
tion twice, the ſecond, or the double of the ſecond made leſſe by the firſt, ſhall come forth
equall to the third, which was to bee proued.

Propoſ 3.   *Of the Logarithmes of three proportionals, the
double of the ſecond, or middle one, is equall to the
ſumme of the extremes.*

By the ſecond Propoſition next going before, the double of the ſecond, made leſſe by
the firſt, is equall to the third. To both the
equall ſides adde the firſt, and there ſhall a-
riſe the double of the ſecond equall to the
firſt and third, that is, to the ſumme of the
extremes, which was to bee demonſtrated.

Propoſ. 4.   *Of the Logarithmes of foure proportionals, the
ſumme of the ſecond and third, made leſſe by the
firſt, is equall to the fourth.*

Seeing by the firſt Propoſition of the *Logarithmes* of 4 proportionals, the ſecond made
leſſe by the firſt, is equall to the fourth leſſe
by the third : adde the third to both ſides of
the equality, and the ſecond and third made
leſſe by the firſt, ſhall bee made equall to the
fourth, which was propounded.

Propoſ. 5.   *Of the Logarithmes of foure proportionals, the
ſumme of the middle ones, that is, of the ſecond
and third, is equall to the Logarithme of the extreames, that is to ſay, the firſt and fourth.*

By the 4 propoſition going afore the 2 &
third

Chap.3.   *The firſt Booke.*   9

third made leſſe by the firſt, were equall to
the fourth : to both ſides of the equality adde
the firſt, and the ſecond more by the third
ſhall bee made equall to the fourth, more by
the firſt, which was to be demonſtrated.

Of the Logarithmes of foure continuall pro-   Propoſ. 6.
portionals, the triple of either of the middle ones, is
equall to the ſumme of the further extreame, and
the double of the nearer.

By the ſecond propoſition, the double of
the ſecond made leſſe by the firſt, is equall to
the third; and by the third propoſition the
double of this, that is, the fourefold
of the ſecond made leſſe by the double of the
firſt, ſhall be equall to the ſumme of his extreames, that is, the fourth more by the ſecond. Now, if from both ſides of the equality
you ſubtract the ſecond, the triple of the ſecond made leſſe by the double of the firſt,
ſhall be made equall to the fourth. Againe,
to the ſides of this equality adde the double
of the firſt, and there ſhall ariſe the triple of
the ſecond, equall to the fourth, more by the
double of the firſt, which wee vndertooke to
proue.

An Admonition.

H Itherto we haue ſhewed the making and
ſymptomes of *Logarithmes;* Now by what
kinde of account or method of calculating
they may be had, it ſhould here bee ſhewed.
But becauſe we do here ſet down the whole
Tables, and all his *Logarithmes* with their
Sines to euery minute of the quadrant : therfore paſſing ouer the doctrine of making
Logarithmes, til a fitter time, we make haſte to
the vſe of them : that the vſe and profit of the
B 5   thing

thing being firſt conceiued, the reſt may
pleaſe the more, being ſet forth hereafter, or
elſe diſpleaſe the leſſe, being buried in ſi-
lence. For I expect the iudgement and cen-
ſure of learned men hereupon, before the
reſt raſhly publiſhed, be expoſed to the detra-
ction of the enuious.

CHAP. III.

*Containing the deſcription of the Table of
Logarithmes, and of the ſeuen
columnes thereof.*

1 Section.

HE *firſt column is expreſſly
of the Arches increaſing from
o to 45 degrees, and is alſo
vnderſtood to bee of their remainders to a ſemicircle.*

2 Section.

*The ſeuenth columne is of
arches decreaſing from a qua-
drant to 45 degrees, and is alſo vnderſtood to bee
of their remainders to a ſemicircle.*

3 Section.

*So the Arches of the one columne are the com-
plements of the Arches of the other anſwering o-
uer againſt them.*

4   *And in the firſt is expreſſed the leſſe ſharpe angle of any right-lined right-angled triangle.*

5   *But in the ſeuenth ouer againſt it, is placed the
greater ſharpe angle of the ſame right-angled triangle.*

6   *In the ſecond columne are the ſines of the arches of the firſt columne.*

7   *And theſe are the leſſe legges ſubtending the
leſſe angle of a right angled triangle, whoſe Baſe,
or Hypotenuſe is the whole ſine.*

8   *In the ſixth columne are the ſines of the arches of the ſeuenth columne.*

9   *And*

CHAP.3. *The firſt Booke.* 11

And theſe are the greater legges ſubtending 9
the greater ſharpe angle of the ſame right-angled
triangle, whoſe Hypotenuſe is the whole ſine.

Hence it followeth, that of the whole ſine, and 10
the ſine of the ſecond columne, and the ſine of the
ſixth columne anſwering ouer-againſt the ſame,
there is made a triangle that is equiangled, and
like to any right-angled right-lined triangle.

The third columne containeth the Logarithmes 11
of the arches and ſines towards the left hand.

Which are the Logarithmes of the proportion 12
of the leſſe legge of a right-angled triangle, to the
Hypotenuſe of the ſame.

And they are alſo the Logarithmes of the com- 13
plements of the arches and ſines towards the right
hand, which we call Antilogarithmes.

The fift columne containeth the Logarithmes 14
of the arches and ſines towards the right hand.

Which are the Logarithmes of the proportion 15
of the greater legge of a right-angled triangle, to
the Hypotenuſe of the ſame.

They are alſo the Antilogarithmes *of the* 16
arches and ſines towards the left hand, or the Lo-
garithmes of the complements.

Laſtly, the fourth or middle columne contai- 17
neth the differences betweene the Logarithmes of
the third and fifth columnes. And ſo this columne
is two-fold, Abounding and Defeĉtiue.

Thoſe differences are Abounding, which ariſe 18
out of the ſubtraĉtion of the Logarithmes of the
fifth columne from the Logarithmes of the third
columne.

But the differences ariſing by ſubtraĉtion of 19
the Logarithmes of the third columne out of the
Logarithmes of the fifth columne, are Defeĉtiue,
which therefore are leſſe then nothing.

The Abounding differences are called the dif- 20
ferentiall

12   *The firſt Booke.*   CHAP.3

ferentiall numbers of the arches towards the left hand.

21   *And are the Logarithmes of the proportion of the lesse legge of a right-angled triangle, to the greater legge of the same.*

22   *And are also the Logarithmes of the Tangents of the left hand arches.*

23   *But the defectiue Differences, are called the differentiall numbers of the right hand arches.*

24   *And are the Logarithmes of the proportion of the greater legge of a right-angled triangle, to the lefte legge of the same.*

25   *And are also the Logarithmes of the Tangents of the right-hand arches.*

26   *Also euery left hand arch, and the remainder thereof to a semicircle, is called the arch of the complement of the arches, sines, & right hand Logarithmes, and of the Defectiue differentials.*

27   *And contrariwise euery right hand arch, and the remainder thereof to a semicircle, is called the arch of the complement of the arches, ſines and left hand Logarithmes, and of the Abounding differentials.*

Admonitions.

28   H*Ere it is to be noted, that if you make the Logarithmes of the third columne Defectiue, ſetting before them this marke, —— they shall bee made the Logarithmes of the Hypotenuſes or Secants of the right hand arches of the seuenth columne.*

29   *And these also shall bee made the Logarithmes of the proportion of the Hypotenuse of a right angled triangle to the lesse legge of the same.*

30   *And if you make the Logarithmes of the fifth columne Defectiue, they shall bee the Logarithmes of the Hypotenuſes, or of the Secants of the left hand*

CHAP.4.   *The firſt Booke.*   13

hand arches of the firſt columne.

The ſame ſhall alſo be the Logarithmes of the 31 *proportion of the Hypotenuſe of a right-angled tri-angle to the greater legge of the ſame. But be-cauſe the ſines onely, and their arches, and the Logarithmes with their Differentials, are ſuffici-ent for attaining the knowledge of right-lined triangles, and for the knowledge of ſphæricall tri-angles, the arches onely with their Logarithmes and Differentials are ſufficient without regard of the ſines. Therefore we haue excluded the Tan-gents, and the Hypotenuſes; or Secants, out of the Table : and in ſphærical triangles we will haue the ſines alſo not regarded; yet we will ſhew you by the way, that you may, if you liſt, vſe them all rea-dily enough in right-lined triangles, but not in ſphæricall.*

CHAP.IV.
Of the vſe of the Table, and of the numbers thereof.

THe Sines, Tangents and Secants Sect. 1. being preciſely found in their Tables, to finde their Loga-rithmes as preciſely.
By the 11 and 14 Section of the third chapter, the Sine giuen being found in the ſe-cond, or ſixth columne of our Table, the Lo-garithme thereof ſhall bee found in the third or fifth columne of the ſame line.

So therefore, the Logarithmes of the Sines that are in the table are exactly had. And the numbers of the Tangents and Secants be-ing found in their owne Tables, you haue their arches.

And

14   *The firſt Booke.*   CHAP.4

And the arches being knowne, our Table
giueth you the *Logarithmes* of the Tangents,
or the differentials with their ſignes or marks
in the middle columne, by the 22 and 25 Sect
Andthe *Logarithmes* of the Secāts reciprocally in the third & fifth columnes; yet ſetting
before them this ſigne ——— by the 28 and
30 Sect. Therefore the *Logarithmes* of the
Sines, Tangents and Secants that are in the
Tables, are thus had.

Examples of Sines.

I Seeke the *Logarithme* of the ſine 694658.
I finde that ſine preciſely in the ſecond columne, anſwering to the arch 44 degrees, o
min. & in the ſame line of the third columne,
there ſtandeth ouer-againſt it, the *Logarithme*
364335 which I ſought. Alſo let the *Logarithme* of the ſine 721357 bee ſought. This
ſine ſhall bee found anſwering to the arch 46
degr. 10 min. and neere adioyning thereto
326620. the Logarithme thereof that was
ſought.

Examples of Tangents.

L Et the *Logarithme* of the Tangent 218645
bee ſought. To this Tangent there anſwereth in his Table the arch of 12 degr. 20
min. and to this arch in the middle columne
of our Table, anſwereth the Logarithme, or
differentiall abounding 1520306 which was
ſought. Alſo if you ſhal ſeeke the Logarithme
of the Tangent 4573629. you ſhall finde in
the Table of Tangents his arch 77 degr.
40 min. and the ſame differentials of this
arch in our Table, but yet defectiue, that is,
——— 1520306.

Examples

CHAP.4.   *The firſt Booke.*   15

Examples of Secants.

TO the Secant 1811801 there anſwereth in the Table of Secants, the arch 56 degr. 30 min. and to this arch in our Table agree- eth reciprocally ——594321 the defeᶜtiue Logarithme of the Secant 1811801, aboue written. So you ſhall find —271425, the Lo- garithme of the Secant 1311834. & of the ſe- cant 1396059. you ſhall find the Logarithme —— 333653.

To æſtimate the Logarithmes of the numbers 2 *giuen, and not found in the Tables of the Sines, Tangents, ond Secants.*

Seeke the number that is moſt like the number giuen in the ſecond or ſixt columne of our Table, whether it be ten fold, an hun- dred fold, a thouſand fold, 10000 fold, 100000 fold, 1000000 fold : or if you will in the Ta- bles of Tangents and Secants : and note the arche hereof. For the Logarithme thereof taken out of our Table, is that you ſeek for : yet keeping in minde, or for memory ſake, ſetting downe in cyphers, the number of the places or figures of the multiplicitie. As if the Logarithme of the number 137 bee ſought, which is not found in the Tables, you ſhall finde among the Sines 1454. 13671. and 137156. And among the Tangents 1370305. but among the Secants, the number 1370505 which is likeſt of all to the number giuen, if the laſt foure figures toward the right hand be vnderſtood to be blotted out. Therefore let the Logarithme of this Secant 1370305. and of his arch 43 degr. 8. min. be ſought out by the former Seᶜtion, or by the 28 and 30 Seᶜtions of the third chapter, and it ſhall bee found——325033, which is alſo taken for the

16   *The firſt Booke.*   CHAP.4

the Logarithme of the number giuen 137 re-
membring, notwithſtanding, that the 4 laſt
figures are to be cut off, or for memory ſake
to be noted thus expreſſly —— 315033 —— 0000
Likewiſe if by the Tangent aboue expreſſed,
1370505 you ſhall ſeeke the Logarithme of
the number 137 by the arch of that Tangent
53 degr. 53. min. ſhall be found by the 25 Section in the middle columne —— 315179,
the Logarithme of that Tangent 1370505
which becauſe it exceedeth 137 the number
giuen by foure places, or figures. Therefore
—— 315179 —— 0000 ſhall be the Logarithme of the number giuen 137 ; yet this
Logarithme is ſo much leſſe exact by how
much 1370505 is more vnlike to the number 1370000, or the 10000 fold of the
number giuen. But this error exceedeth not
$\frac{505}{10000}$. Laſtly, if you ſhall ſeeke the Loga-
garithme of the number giuen 137 by the
Sine aboue written 137156. that ſhall bee
found to bee 1986633 —— 000 by this &
the 11 Section of the third chapter. In like
manner you ſhall work by the ſigne ✛ when
the number of the figures of the quantitie
giuen, exceedeth the number of the figures
of the ſine that is likeſt thereto, which ſel-
dome happeneth. As if the Logarithme of the
number (or diſcreet quantitie) 232702 bee
ſought for, you ſhall finde in the Table, the
ſine 2327 moſt like thereto; but it wanteth
two figurs. Therfore to the Logarithme hereof, found in the Table (by the 11 Sect. chap. 3)
which is, 6063128. let be added two cyphers,
the ſigne ✛ being put betweene, and it ſhall
be made 6063128 ✛ 00 for the Logarithme
of the number 232702. which was ſought for.

                                But

CHAP.4.   *The firſt Booke.*   17

But the beſt way of eſtimating Logarithmes, is that whereby they were firſt made,
wherof we ſhall ſpeake in another place.

Therfore as in the firſt Section going afore, ſimple 3
and pure Logarithmes are giuen : ſo in this Secti-
on next going before by putting cyphers to them,
they become impure.

To adde Logarithmes of like ſignes, is to giue the 4
ſumme of them both, with their ſigne common to
them both.

As by the Addition of —— 56312 to
—— 73495. there ſhall come forth —— 129807.
Alſo 4216 being added to + 5392, there
comes forth 9608. So 3219 —— 00 added to
4360 —— 000 make 7579 —— 00000.

To adde the Logarithmes of vnlike ſignes, is to 5
giue the difference of them with the ſigne of the
greater number.

As of the Addition of —— 210 to 332 is prodnced + 122.

Alſo of the addition of —— 210 to 192, comes
forth —— 18.

So —— 210 + 000 added to 332 —— 00 are
122 + o.

Alſo —— 210 —— 000 added to 192 + 00,
are —— 18 —— o.

Of two Logarithmes this is properly ſaid to bee 6
the Defective of that, and that the Abounding of
this : when they haue both number and cyphers
common, or the ſame, and all the ſignes + and
—— altogether contrary.

As of the Abounding Logarithme 56312,
the defective is —— 56312. Alſo of the Abounding Logarithme 56312 —— 00 the Defective
is —— 56312 + 00. So of the Abounding
Logarithme 56312 + 00, the Defective is,
56312 —— 00.

18   *The firſt Booke.*   CHAP.4

7   *To ſubtraƈt an abounding Logarithme, is to*
adde his defeƈtiue.

As to ſubtraƈt the abounding Logarithme
56312 out of —— 73495, is the ſame as to adde
his defeƈtiue which (by the ſixth Seƈtion.) is
—— 56312 to the ſame —— 73495, and they
ſhall bee made (by the 4 going before)
—— 129807. So to ſubtraƈt 56312 ╋ oo out
of —— 73495 —— ooo is the ſame as to adde
—— 56312 ╋ o to 73495 —— ooo, and they
are made (by the 4 and 5 Seƈt. going before) —— 129807 —— ooooo.

8   *To ſubtraƈt a defeƈtiue is to adde his aboun-*
dant.

As to ſubtraƈt a defeƈtiue —— 4216 out
of ╋ 5392, is the ſame that it is to adde 4216
to 5392. and (by the fourth Seƈtion) to bring
forth 9608. So it is the ſame to ſubtraƈt
—— 4216 ╋ oo out of 5392 ╋ o, that it is to
adde 4216 —— oo to 5392 ╋ o. and to bring
forth 9608 — o.

9   *To increaſe or diminiſh a Logarithme in*
number, his former value remaining, is to
adde to it, or ſubtraƈt from it, any of the Lo-
garithmes following, as 2302584 ╋ o, *or*
4605168 ╋ oo, *or* 6907753 ╋ ooo, *or*
9210337 ╋ oooo, *or* 11512921 ╋ ooooo,
ſignifying nothing at all.

As let the Logarithme bee 3916 —— o
whereto if you adde any of them, as for example ſake, 2302584 ╋ o, there ſhall bee
made thereof 2306500 greater in number,
but in value altogether the ſame that
3916 —— o is : for the quantity or numerall
value of this Logarithme 3916 —— o (by the
12 and 13 Seƈtions following of this Chapter) is 996092, from which take onely the
laſt

CHAP.4. *The firſt Booke.* 19

laſt figure, as —— o, ſignifieth, and it ſhall be made 99609. And the numeral value of that Logarithme 2306500 (by the 12 an 13 Se-ctions following of this chapter) is alſo 99609 the ſame that was before.

An example of diminiſhing.

Let the Logarithme 2545177 bee to be di-miniſhed , from which if you ſubtract 2302584 + o, there is left 242593 —— o of the ſame value that this former 2545177 was. For the value of the ſimple and pure Logarithme 242593 is ten fold the value of either of them. Their values therefore are e-quall each to other. For the addition of the Logarithme 2302584 + o, ſignifieth nothing elſe, but that the value of the number where-to it is added, is to be diuided into ten parts, and that one cypher is to bee added to this tenth part : but the ſubtraction of the ſame ſignifieth that the value of the Logarithme from whence it is ſubtracted, is made tenne fold more, and that one cypher is caſt away from this ten fold. There remaineth therfore the ſame value in both of them.

So 46051684 + oo added, ſignifieth that two cyphers are added to the hundreth part of the value : and being ſubtracted, it ſigni-fieth that two cyphers are caſt away from the hundreth fold, and ſo of the reſt aboue ex-preſſed.

An Admonition.

B*Vt becauſe the addition and ſubtraction of theſe former numbers may ſeeme ſomewhat painfull, I intend (if it ſhall pleaſe God) in a ſe-cond Edition, to ſet out ſuch Logarithmes as ſhal make thoſe numbers aboue written to fall vpon decimal*

20   *The firſt Booke.*   CHAP.4

decimal numbers, ſuch as 100,000,000, 200,000,
000, 300,000,000, *&c. which are eaſie to bee
added or abated to or from any other number.*

10   *If therefore you ſhall adde to a* Logarithme
*that is diminiſhed by ſome cyphers, or ſhall
ſubtract from a* Logarithme *increaſed by cyphers, any of the* Logarithmes *aboue written that
containe ſo many cyphers, there ſhall out of an impure* Logarithme *bee produced, a pure one of the
ſame value.*

As in the firſt example going before, let the
impure Logarithme 3916 —— o bee to bee
purged from his cypher and ſigne ——, adde
therefore thereto 2302584 $+$ o there ſhall
thereof be made, as before, 2306500, the
pure Logarithme of his former value. So
from the impure Logarithme 6358447 $+$ oo
if you ſubtract 4605168$+$oo, (which containeth as many cyphers) there ſhall remaine
the pure Logarithme 1753278, and of the
ſame value, whereof that former impure Logarithme was.

11   *If to a* Logarithme *that is Defectiue in number, you ſhall adde any of the foreſaid* Logarithmes *of the ninth Section, that is greater in number, there ſhall come forth a* Logarithme *of the
ſame value Abounding in number.*

As to the Logarithme —— 2859527 —— oooo
adde any of the numbers of the ninth Section, that is greater in number. As for example, 4605168 $+$ oo, and there ſhall bee made
thereof 1745641 —— oo of the ſame value,
and Abounding in number.

12   *You may giue the Sines, Tangents, and Secants, or any numerall values whatſoeuer, of the*
Logarithmes *that are found in our Table by the*
11. 14. 22. 25. 28. 30. *Section of the 3 Chapter,*
                                              whether

CHAP.5.   *The firſt Booke.*   21

whether they be pure or impure.

As to the Logarithme of 36 degrees, 40 minutes 515572, in the third columne, anſwereth his ſine 597159 in the ſecond columne, & to the Defectiue therof —— 515572 there anſwereth in the Table of Secants, 1674597 , the Secant of 53 degrees, 20 minutes.

Alſo to the Differentiall Logarithme 295079 in the fourth columne, anſwereth the Tangent 744472 in his Table, and to his Defectiue —— 295079 anſwereth 1343233 the Tangent of 53 degrees and 20 minutes. So of the Logarithme 220493 in the fifth columne, the numerall value in the ſixth columne is 802123, that is the Sine of 53 deg. and 20 min and the numerall value of the Defectiue thereof, that is —— 220493 is the Secant 1246691, agreeing to 36 degrees and 40 minutes.

An example of impure Logarithmes.

L Et the value of the impure Logarithme 9780 — o bee to bee ſought out; to this number, there anſwereth in our Table the Sine 990268, from which take the figure next the right hand (as —— o doth ſhew) & they ſhall be made 99027, the value of the Logarithme 9780 —o which was ſought. So the value of the Logarithme 2545177 + oo is 7845900, becauſe that to the pure Logarithme 2545177 there anſwereth in our table the Sine 78459. Alſo of the Logarithme 34914 —— oo found in the fourth columne at 46 degrees, the value ſhall be 10355, becauſe the Tangent of 46 degr. is 1035530. So of the Logarithme —— 635030 ——oo found in the third columne at 32 degrees, the value

value is 18871 , becauſe the Secant of the complement of 32 degrees, that is of 58 degrees, is 1887080, whoſe two laſt figures next the right hand 80, are to be blotted out for —— 00 adioyned to the Logarithme.

13   *To eſtimate the numerall values of the* Logarithmes *giuen, and not found in our Table.*

For common meaſuring, it is ſufficient for the moſt part, to take for the Logarithme giuen, the numeral value of the Logarithme in the Table, that comes neereſt that, which is giuen. But if you deſire to come neerer the marke, increaſe or diminiſh in number the Logarithme giuen, by the 9 Section of this chapter, his former value remaining vntill it be either found in the Table, or become like enough to ſome Logarithme in the Table, and the value of this Logarithme found by the former Section, is that which we ſeek for. As for example, let the value of this Logarithme 2314972 ✛ 0 bee ſought, to which there is none found like or neere enough in the Table; but if you ſubtract from it 2302584✛0, there ſhal be left 12388 almoſt, to which vnder 81 degr. there ſhall be found one that is neere, and like enough to it, that is, 12388, the Sine whereof 987688 found by the former Section, is the value of the Logarithme propoſed 2314972 ✛ 0 which was ſought for.

An Admonition.

FOr this and the 2 Sect. of this chapter, we would haue you admoniſhed, that the Logarithmes of the numbers giuen, & contrariwiſe the numerall values of the Logarithmes giuen, when they are not found in the Table, are moſt exactly giuen by the way, by which

CHAP.5.   *The firſt Booke.*   23

which the Logarithmes are made or reſolued, which is that you deſcend from the ſine
giuen by meanes Geometrically proportionall, vntill you come to the next leſſe ſine in
the Table. Likewiſe from the Logarithme
heereof, in the Table, that you deſcend
alſo by as many agreeable meanes Arithmeticall : and the laſt of theſe ſhall be the Logarithme of the firſt of them, and contrariwiſe
by reſolution that you deſcend from the Logarithme giuen by Arithmeticaall meanes to
the next leſſe Logarithme in the Table, and
from the value of this in the Table likewiſe,
that you deſcend, by as many meanes Geometricall and agreeble : and the laſt of theſe
ſhall bee the numerall value of the firſt of
thoſe Logarithmes. But what Arithmeticall
equalitie of difference agreeth and is fitting
to euery continued Geometricall proportion, is a matter of no meane skil to finde out.
Wherefore of theſe (if God will) we ſhall in-
treate hereafter more at large, when we ſhall
handle the making of Logarithmes.

CHAP.V.

Of the moſt ample vſe of the Logarithmes,
and ready praᴄtiſe by them.

F the Logarithmes of three proportio-   *1. Probleme*
nals, the middle Logarithme being gi-
uen, and one extreame to finde the o-
ther extreame, or his proportionall, or
arch by one doubling, or ſubtraᴄtion onely.
  Seeing that by the ſecond propoſition,
Chap.2. the double of the middle (Logarithme) made leſſe by one of the extreames,
is made equall to the other; Therefore from
the

24   *The firſt Booke.*   CHAP.4

the double of the middle Logarithme giuen,
ſubtract the giuen Logarithme of the extreame, and there ſhall remaine the Logarithme of the extreame that was ſought for :
which being found in the third, fourth, or
fifth columne of the Table, you haue the arch
anſwering thereto in the firſt and ſeuenth columne, and the Sine in the ſecond or ſixth,
and their Secants or Tangents in their Tables, by the third Chapter, Section 1. 2. 6. 8.
11. 14. 22. 25. 28. 30. for the extreame that
was ſought for. *Example.*

L ET the firſt proportionall giuen, bee
1000000, and the ſecond 707107 : let the
third be ſought for, which commonly is found
by multiplying the middle number by it ſelfe,
& diuiding this ſquare by the firſt. But we find
it eaſilier by doubling the Log : of the middle
number 346573, and by ſubtracting from this
double (wᶜʰ is 693147) the Logarithme of the
firſt, which is 0, & there remaineth 693147,
the Logarithme ſought for, whoſe arch you
ſhall finde to be 30 degrees, and the Sine adioyning thereto 500000, which is the proportionall number ſought for. Therefore
1000000. 707107. 500000, are three proportionall numbers, the laſt whereof wee
found onely by doubling, and ſubtraction,
which wee promiſed. Alſo let there bee two
proportionall numbers giuen, the firſt
1056256, & 766045 the ſecond, or at leaſt
their Logarithmes ——54730, and 266515.
The third you ſhall thus finde : From
the double of this laſt 533030 ſubtract
—— 54730, and by the 8 Section of the 4
chapter, there is brought forth 587760, the
Logarithme of 33 degrees. 45 minutes, the
                                            ſine

CHAP.5.   *The firſt Booke.*   25

ſine whereof 555570 is the third proportionall number that was ſought for.

Out of the Logarithmes of three proportionals,   *Prob.* 2.
*the extreame Logarithmes being giuen, to finde
the middle Logarithme and his proportionall and
arch, by one addition onely, and diuiſion by two.*

Seeing by the third propoſition of the ſecond chapter, the double of the middle Logarithme is equall to the ſumme of the extreames, therefore adde the Log. of the extremes, and diuide the product by 2, & there
ſhall come forth the Logarithme of the middle proportionall number : and thereby the
middle proportionall, and the arch thereof,
is knowne in the columnes, and by the Sections, as before. *As for Example.*

L Et the extremes 1000000 and 500000
bee giuen, and let the meane proportionall be ſought : that commonly is found by
multiplying the extreames giuen, one by a-
nother, and extracting the ſquare root of the
product. But we finde it eaſilier thus; We
adde the Logarithmes of the extreames o
and 693147, the ſumme whereof is 693147
which we diuide by 2 & the quotient 346573
ſhall be the Logar. of the middle proportionall deſired. By which the middle proportionall 707107, and his arch 45 degrees are
found as before. Alſo let the extremes giuen
bee 1056256 and 555570, their Logarithmes are ——54730 and 587760. The
ſumme of theſe put together, is 533030 by
the 5 Sect. Chap.4. which we diuide by two,
and the quotient is 266515, the Logarithme
and his arch 50 degr. and the ſine or meane
proportionall ſought for is 766044. found
by addition onely, and diuiſion by two.

C   *Out*

Prob. 3.

Out of the Logarithmes of foure proportionals, three being giuen, or their arches, to find the fourth Logarithme with the ſine and arch thereof, by one addition onely, and ſubtraction.

In this probleme wee alwayes make the thing demanded the fourth, ſo that as the firſt of the numbers giuen is to the ſecond, ſo is the third to the number demanded. And ſeeing the ſumme of the Logarithmes of the ſecond and third of the numbers ſo placed, diminiſhed by the Logarithmes of the firſt, is equall to the Logarithme of the fourth, by the 4. Prop. Chap. 2. Therefore adde the Logarithmes of the ſecond and third, and from the ſumme of them take the Logarithme of the firſt, and there ſhall remaine the Logarithme of the fourth proportionall number demanded, and thence the fourth number it ſelfe, and the arch thereof.

For examples ſake.

AS 766044 is to 984808 : ſo let 500000 be to the fourth proportionall which wee ſeeke for. This they commonly finde by multiplying the ſecond and third, and diuiding the product by the firſt. But you may find it more eaſily thus : Adde the Logarithme of the ſecond 15309, and of the third 693147, the ſumme whereof ſhall be 708456 : out of which ſubtract the Logarithme of the firſt, which is, 266515, and there ſhall remaine 441941, the Logarithme of the fourth, whoſe ſine 642788 is the fourth proportionall deſired, and the arch thereof 40 degrees. The ſame would come forth if (the ſines being neglected) their three arches onely were giuen 50 degrees, 80 degrees, and 30 degrees. For out of the Logarithmes of the arches 80 degrees,

CHAP.5. *The firſt Booke.* 27

degrees, and 30 degrees, ſubtract the Logarithme of 50 degrees, there ſhall remaine the Logarithme of 40 degrees : and ſo the arch it ſelfe 40 degrees, ſhall be knowne without the ſines, or their multiplication or diuiſion, according as we promiſed in the beginning.

Another Example.

AS the Tangent of 43 degrees is to the Sine of 57 degrees, ſo let the Tangent of 35 degrees bee to a fourth Sine vnknowne, whoſe arch without regard either of Sines or Tangents, we ſhall thus finde : Wee adde the Differential Logarithme of 35 degrees, that is, 356378 found in the middle columne to the Logarithme of 57 degr. that is 175937 placed in the fifth columne from the product, that is, 532316, wee Subtract the Differentiall of 43 degrees, which is 69870, and there remaineth 462446, the Logarithme of the fourth (Sine) which being found in the third columne, by the 11 Section of the third chapter, you ſhall finde cloſe by it in the firſt columne 39 degrees 2 minutes almoſt, which is the arch of the fourth proportionall, or Sine neglected.

Thus the arches of proportionall numbers are found without their Sines, Tangents, Secants, or any proportionall numbers whatſoeuer.

Which ſo ſhort a way of working, doth helpe very much for meaſuring the angles of plaine triangles, and for the whole *Trigonometrie* of ſphærical triangles, as in his proper placed ſhall appeare.

Of foure numbers in continuall proportion, the Prob. 4. *extremes being giuen, or their arches, to finde any*

C 2 *of*

*of the middle numbers, or any of their arches,
onely by diuiding by three, inſtead of the troubleſome extraɛting of the cubicke root.*

Seeing that in the Logarithmes of theſe, the triple of any middle one, is equall to the ſumme of the extreme remoued, and the double of the next extreame, by the ſixth propoſition of the ſecond chapter. Therefore adde the double of either extreame Logarithme to the Logarithme of the extreame remaining, and diuide the product by three, and there ſhall come forth the Logarithme of the middle proportionall next the former extreame, and after the ſame manner, the other meane proportionall alſo.

As for examples ſake.

LEt the firſt extreame be 402925, and the laſt, 1056256, the meane proportionals are ſought for, which without extraction of the cubicke roote you ſhall thus finde. The Logarithme of the numbers giuen are 909005, and —— 54730: to the double of that 1818010, adde this, and the ſumme ſhall bee 1763280, which diuided by three, bringeth forth 587760 the Logarithme, whoſe Sine 555570 is the firſt meane proportionall ſought for. Alſo in like manner to the double of this —— 547305 which is —— 109460, adde that 909005, and the product will bee 799545, which diuided by three, bringeth forth 266515 the Logarithme, whoſe Sine 766044 is the later meane which was ſought for. Theſe therefore are foure continuall proportionals 402925, 555570, 766044 and 1056256.

Another example.

Let the extreames giuen bee 1414213,
and

CHAP.5   *The firſt Booke.*   29

and 500000 : the firſt of theſe being found
in the Table of Secants, the Logarithme
thereof in our Table is——346573, and the
Logarithme of 500000 is 693147 to the
double whereof, 1386294 adde ——346573,
the ſumme ſhall be 1039720, which diuided
by 3, will be 346573 the Logarithme of the
meane proportional next the leſſe extreme,
which is 707107. So to the double of
——346573, which is ——+ 693147,
add 693147, and there ſhall be made thereof
nothing, which diuided by 3, maketh alſo o,
the ſine and the value whereof is 1000000
for the remaining and greater meane proportionall. Theſe foure therefore are continually proportionall, 1414213. 1000000,
707107. 500000.

The Concluſion.

NOw out of this that is already deliuered,
let the learned iudge how great benefit
the *Logarithmes* bring them; ſeeing that by
the addition and ſubtraction of them, and by
diuiding by 2 and 3. and by other eaſie addi-
tions, or ſubtractions, multiplication, diuiſion : the extraction of the ſquare and cubicke
rootes, and all the great toyle of calculating
is auoided, a generall taſte whereof we haue
giuen in this Booke.

But in the booke following we ſhall treate
of their proper and particular vſe in that noble kinde of *Geometrie* which is called *Triga-
nometrie.*

The end of the firſt Booke.

C 3   *The*

30

THE SECOND
BOOKE.

---

*Of the excellent vse of the admirable Ta-
ble of Logarithmes in Trigonometrie.*
CHAP. I.

EEING that *Geometrie*, is
the Art of meaſuring well,
and meaſuring belongeth
to Magnitudes, and Magni-
tudes are Figures, (at leaſt
in power) and a Figure is
either a Triangle, or Tri-
angled, and that which is triangled, is com-
pounded, or made of Triangles : which, and
whoſe parts, being meaſured, that figure alſo, and all the parts thereof will bee meaſured. It is therefore certaine, that the Arithmetical ſolution of any Geometricall queſtion, dependeth on the doctrine of Triangles.
A Triangle is either right-lined or ſphærical.

Of Right-lined Triangles.

Propoſ. 1.   *The three Angles of a right-lined Triangle, are
equall to two right Angles.*
       Therefore if two angles be giuen, take the
                                        ſumme

CHAP.1.   *The fecond Booke.*   31

fumme of them out of 180 degrees, and there
will come forth the third angle. Alfo one angle being taken out of 180 degr. the fumme
of the other two remaines.

A Right-lined Triangle is either right-
angled, or obliquangled. In right angled triangles wee call the fides that are about the
right angle, Leggs; and that which fubtendeth the right angle wee call the *Hypotenufe.*

In a right-angled triangle the Logarithme of the   *Propof. 2.*
legg, is equall to both the Logarithme of the angle
oppofite thereto, and the Logarithme of the Hypo-
tenufe.

Seeing it is manifeft out of the princi-
ples of *Trigonometrie,* that either legg hath
the fame proportion to the fine of the angle
oppofite therto, that the *Hypotenufe* hath to
the whole fine : and (by the fifth propofition
of the fecond chapter of the 1 book) the Logarithmes of the fecond and third of thefe
foure proportionals are equall to the Logarithmes of the firft and fourth : and the Logarithme of the fourth is 0, or nothing, by the
Corolarie of the fixth definition of the firft
Chapter, and firft Booke.

Therefore (as before) the Logarithme of
the legge is equall to the fumme of the Logarithme of the angle which it fubtendeth, and the Logarithme of the Hypotenufe.

Therefore any two of the Hypotenufe,   *A confe-*
the legg, and angle which it fubtendeth,   *quent.*
being giuen : the third, and thence all the reft
of the parts of a right angled triangle will bee
knowne.

Becaufe thefe three, with the whole fine,

C 4   make

32   *The ſecond Booke.*   CHAP.I

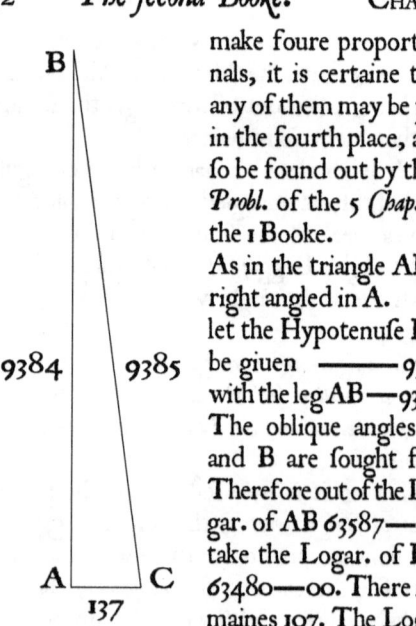

B

9384   9385

A   C

137

make foure proportio-
nals, it is certaine that
any of them may be put
in the fourth place, and
ſo be found out by the 3
Probl. of the 5 *Chap.* of
the 1 Booke.
As in the triangle ABC
right angled in A.
let the Hypotenuſe BC
be giuen ———9385
with the leg AB—9384
The oblique angles C
and B are ſought for.
Therefore out of the Lo-
gar. of AB 63587—00
take the Logar. of BC
63480—00. There re-
maines 107. The Loga-
rithme of the Angle C,
whereto there anſwereth in the Table 89
degrees, 9 $\frac{3}{4}$ for the angle C, and oueragainſt
it o degr. 50 $\frac{1}{4}$ for the complement therof, namely the Angle B.

Contrariwiſe, if the Angle C bee giuen,
with the legge of the right angle AB, and the
Hypotenuſe BC be ſought for.

Out of the Logar. of AB. + 63587 — 00
Take the Logar. of the angle C + 107
_____

And there will come forth— + 63480 — 00
the Logarithme of BC 9385 the Hyponenuſe
that was ſought for.

Thirdly, if BC and the Angle C being giuen, and AB be ſought for,
Adde the Logarithme of BC + 63480—00
to the Logar. of the angle C + 107

And

CHAP.I.   *The second Booke.*   33

And there will be brought forth 63587—00 the Logarithme of the number 9384, anſwering to the legge AB, which was ſought for.

No otherwiſe is the legge remaining AC found by the angle B. (which is the complement of the angle C) already knowne. And ſo all the parts of this right-angled triangle are knowne.

In a right angled triangle the Logarithme of   Propoſ. 3. *any legge is equall to the ſumme of the Differentiall of the oppoſite angle, and the Logarithme of the leg remaining.*

Seeing it is manifeſt out of the common doctrine of Triangles, that either legge hath the ſame proportion to the Tangent of the angle oppoſite thereto, that the other hath to the whole ſine : and ſeeing that (by the 5 propoſition of the ſecond chapter of the firſt Booke) of theſe foure proportionals, the Logarithmes of the middle ones, that is to ſay, the Differential of the angle, and the Logarithme of the legge including it, are equall to the Logarithmes of the legge ſubtending the ſame, and of the whole ſine, (which is o, or nothing) therefore the Logarithme of the legge is equall to the ſumme, &c. as before.

Therefore of the legges of the right angle, and   Corolarie. *the angle oppoſite to one of them, any two being giuen, the third is knowne (by this Prop.) and therfore all the other parts of the right angled triangle by the former propoſition.*

Becauſe theſe three, with the whole ſine, doe make foure proportionals, it is certaine that euery one of them may be placed in the fourth place, and bee found out by the third *Prob.* of the 5 *Chap.* of the firſt Booke.

As

34   *The ſecond Booke.*   Chap.1

As in the triangle going before **ABC**, right angled at **A** : the leggs **AB** being giuen 9384, and **AC** 137, let the angle **B** be ſought out.
From the Logarithme **AC** $+$ 4292453 — 00
Subtract the Logar. of **AB**, $+$ 63587 — 00
And there will come forth $+$ 4228866   the Differentiall of the angle **B**, o deg. 56. íi. which was ſought for.

But if the legg **AC** be giuen 137, and the angle **B**, o degr. 56. íi, the legg **AB**, ſhall be thus found.
Out of the *Logarithme* of **AC** $+$ 4292453 — 00
Subtract the Differential $\big\}$ $+$ 4228866 — 00
    of the Angle **B**
the number comming thereof $+$ 63587 — 00
is the Logarithme of the number 9384 which is the legg ſought for **AB**.

Thirdly, the legg **AB** being giuen 9384, and the angle **B**, o deg. 56. íi. that the legg **AC** may be found,
Adde the Logar. of the leg **AB** $+$ 63587 — 00
to the Differ. of the angle **B** $+$ 4228866
and there will come forth $+$ 4292453 — 00
the Logarithme of 137 the legg **AC**, which was ſought for.

The Hypotenuſe **BC** is found by the former propoſition. Alſo the angle **C** is knowne, becauſe it is the complement of the angle **B**, already knowne. And ſo by this, and the former propoſition, by any ſide, and any other part of a right-angled triangle giuen, all the other parts thereof are made knowne.

You haue therfore the knowledge of right-angled right-lined triangles accompliſhed : Now of oblique angled triangles.

Chap.

CHAP.2.   *The fecond Booke.*   35

*Of Right-lined Triangles, efpecially
obliquangled.*
CHAP. II.

N *any Triangle : the fumme of*   *Propof.* 4.
*the Logarithmes of any angle
and fide inclofing the fame, is
equall to the fumme of the Logarithmes of the fide, and the
angle oppofite to them.*
  Becaufe, *there is the fame
proportion of all the fides to the fines of the oppofite
angles* : and fo the product of the right fine
of any angle, & any fide including the fame,
is equall to the product of the fide fubtending the former angle, and the fine of the angle fubtended by the firft fide. Therefore by
the *5.Prop. 2.Chap. 1.Booke.* the fumme of the
Lograrithme, &c. is equall, as before.
  Therefore, of two angles whatfoeuer of a kinde   *Corolarie.*
*giuen, and their fubtendants : if three be giuen, a-
ny fourth will be knowne, and thence all the other
parts of the triangle.*
  For of thefe foure proportionals, any that is
fought for, may be put in the fourth place, &
be found out by the third *Prob. Chap.5. Book 1.*
  As of the obliquangled Triangle ABC, let
AB be giuen 26302, and BC 57955, and the
angle C 26 degrees : and let the angle A be
fought for, which is thus found.
Adde the Logarith. of BC +545471—0
To the Logar.of C.26.deg.+824689
And there will bee made   + 1370160—0
From thence take the⎱   + 1335492 — 0
Logarithme of AB, ⎰
  There remaines————34688— 0 the
         Loga-

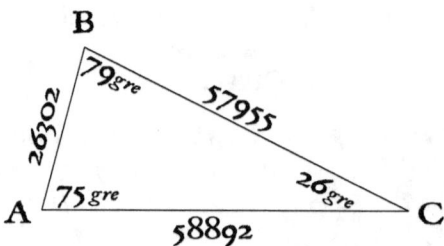

Logarithme of 75 degrees, and a little more wᶜʰ is the angle A ſought for if A appeare to be an acute angle, otherwiſe 105 deg. (by the 1 and 2 Sect. chap.3. book.1.) if it appeare to be an obtuſe angle.

Contrariwiſe, if the angle A bee giuen 75 degr. and the angle C, and the ſide BC as before, and AB be ſought for.

Adde the Logarithme of BC $+$545471—0
to the Logar. of the angle C $+$ 824689

they will made as afore $+$ 1370160—0
From which take the ⎫
Logar. of the angle A ⎰ $-+$ 34668

There will come forth $+$ 1335492—0 the Logarithme of the ſide AB, and the number thereof 26302, which was ſought for.

The angles A 75 degr. and C 26 deg. being now found, the angle B ſhal be 79 deg (by the *1.Prop.* of this book:) out of which being now found, the ſide oppoſite thereto AC 58892 is no otherwiſe found then the ſide oppoſite thereto (AB) was lately found by the angle C. Therfore now all the parts of this oblique-angled triangle are knowne.

In the obliquangled triangles, we call them legs
which

CHAP.2. *The second Booke.* 37

*which are about any angle, & the base which sub-
tendeth the same.*

In obliquangled triangles, the Logarithme of the *Propos. 5.*
*summe of the legges, subtracted from the summe
made of the Logarithme of the difference of the
legs, and the Differentiall of halfe the summe of his
oppofite angles, leaueth the Differentiall of halfe
the difference of the same.*

Becaufe *as the summe of the legges is to the difference of the legges ; so is the Tangent of halfe the
summe of their oppofite angles to the Tangent of
halfe the difference of the same :* Therfore they
are proportionall, and by the *1 Prop. 2.Chap.
1 Book.* the differences, or exceffes of their
Logarithmes are equall. Therfore (by the *4.
Prop. 2.chap. 1.book*) we muft neceffarily conclude as before.

Therefore by two legs, and the angle contai- *A Corolarie*
*ned betweene them, are knowne by this Propofiti-
on, the other oppofite angles, and thereby the o-
ther side, by the propofition going before.*

For the Logarithme of the fumme of the
leggs being fubducted out of the fumme
made of the Logarithme of the difference of
the leggs, and the Differentiall of halfe
the fumme of the oppofite angles put together, there fhall come forth the Differentiall of halfe the difference of the fame angles; which halfe difference being added
to the halfe fumme aforefaid, there fhall
come forth the greater angle ; and being
fubtracted, the leffe.

As in the forefaid *Obliquangled triangle* ABC
Let there be giuen AB one legg   26302
BC th'other leg   57955
B the angle contained
betweene them, 79 degrees, and let the other
angles

angles A and C be ſought for.

The ſumme of the legges AB, and BC is 84257, the Logar. thereof 2473882, and the difference of the ſame AB and BC is 31653 the Logar. thereof 3452921. And ſeeing the angle B is giuen 79 deg. (by the firſt of this book) the ſumme of the angles A and C will be 101 degr. and halfe the ſumme 50 deg. 30. the Differentiall whereof is ——— 193177 Which being added to the Loga-⌉
rithme of the difference of AB, } 3452921
and BC——— ——— ——— ⌋

There wil ariſe ——— ——— $+$3259744
Out of which ſubtract the Loga-⌉
rithme of the ſumme of the legs } 2473882
AB and BC ——— ——— ⌋

Theſe will remaine ——— $+$785862 the Differential of 24 deg. 30, which are the halfe-difference, of the angles A and C that were ſought for.

Therfore adde this half-differ. 24 deg. 30. to the halfe ſumme ——— 50 deg. 30.

And they will make ——— 75 deg. 0. for the angle A, the greater of the angles ſought for.

And out of the——————50 deg. 30. Subtract the ſame ———24 deg. 30.

And there will remaine —— 26 deg. 0. for the angle C, the leſſe of the angles ſought for.

A Definitiõ   *In obliquangled triangles, the true baſe is alwayes, either the ſum of the caſes (& then the difference of the caſes is called the Alterne baſe:) or
the true baſe is the difference of the caſes, & then
we call the ſumme of the caſes, the Alterne baſe.*

As in the Triangle ABC.

The

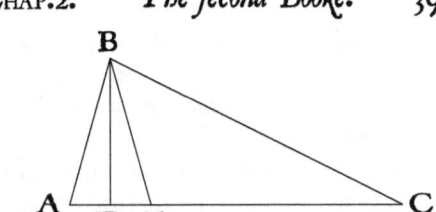

The leſſe caſe is ——AD
The greater caſe is —DC
The ſum of the caſes AC, is the true baſe
And in this triangle take the leſſe caſe AD,
    or that which is equall to it————DE,
    from the greater caſe ————————DC,
    and there will remaine ————————EC,
the difference of the caſes which we call the
Alterne baſe.

Contrariwiſe, in the triangle EBC.
The leſſe caſe is————————DE (Whereto
    AD is equall)
The greater caſe is————————DC
The difference of the caſes EC is the true
baſe
But the ſum of the caſes, that is, AC, wee call
the *Alterne baſe.*

In obliquangled triangles, the ſumme of the Lo-
garithmes of the ſumme, and difference of the legs   Propoſ. 6.
is equall to the ſumme of the Logarithmes of the
true and Alterne baſe.

Becauſe the true baſe is to the ſumme of the
leggs, as the difference of the leggs is to the
Alterne baſe; therfore (by the *5.Prop .2.Chap. 1.*
Book.) we neceſſarily conclude, that the Logarithmes of the baſes are equall to the Logarithmes of the ſumme, and the difference of the
leggs, as before.
Therfore, of an obliquangled triangle, whoſe ſides   *A Corolarie*
are giue, 2 right angled triangles are made, whoſe
                    Hypo-

40   *The second Booke.*   CHAP.2

Hypotenuſes are knowne with one of the legges
of either of them, which (by the ſecond of this book)
makes knowne all the other parts of the obliquan-
gled triangle.

For the Logarithme of the ſumme of the
leggs being added to the Logarithme of the
difference of the leggs, and the Logarithme
of the true baſe being taken from hence,
there will come forth the Logarithme of the
Alterne baſe (by the *4.Prop.* of the *2 Chap.* and
3 Problem of the *5 Chap.* of the *1 Book.*) Therfore halfe the ſumme of theſe baſes is the
greater caſe, and the halfe-difference is the
leſſe caſe.

As in the former triangle ABC,
Let the ſides be giuen, that is to ſay,
The legge   AB   26302
The legge   BC   57955
and the true baſe AC 58892 and let the reſt
bee ſought for.

The ſumme of the leggs is, 84257
     the Logarithme therof is —— +2473882
The diffrence of the legs is, 31653
and the Logar: thereof is, —— +3452921
Adde theſe Logarithmes to- ⎤
gether, and they will make, ⎦ —+ 5926803
From wᶜʰ take the Log: of   ⎫ —+ 2831930
the true baſe AC ———— ⎭
There remaines ————————+3094873
the Logarithme of the number of the Al-
     terne baſe EC ————————45286
which adde to the num- ⎤
ber of the true baſe AC. ⎦ —— 58892
And there remaines   ———— 104178
The halfe whereof DC ———— 52089
is the greater caſe.
Subtract one out of another, viz.

                                        Out

CHAP.2.　　*The second Booke.*　　41

Out of the true bafe AC 58892

Take the Alterne bafe EC 45286

And there remaines —— 13606

The halfe whereof AD.⎫ 6803
is the leffe cafe. ——⎭

Therefore of the right-angled triangle A DB, the Hypotenufe AB, and one of the legs AD being found : and of the right angled triangle BDC the Hypotenufe BC, and the leg DC being found (by the fecond of this chapter) the angles of the right angled triangle at A, and B, and C, are known, and by by confequent, alfo all the parts of the obliquangled triangle propofed, are manifefted by the premifes. Neither fhould you doe otherwife if the fides of the triangle EBC, were giuen, and the other parts were fought. For out of the legges, and the true bafe EC, the Alterne bafe AC is knowne, and out of thefe both cafes, and the reft, as before.

The Conclufion.

NOW therefore, you haue the doctrine of all right-lined triangles perfected & accomplifhed, which if it feeme fomewhat toylefome in finding out the Logarithmes of variable right-lines; yet in calculating the motions of the planets, (in which the excentricities of the Orbs, the diftances of the Auges & Apogæs the diameters of the Epicycles and other right lines, remaine the fame, and invariable) their Logarithmes being once exactly fet downe, fhall alwayes ferue afterwardes without any changing, with maruailous facilitie and certaintie.

Now

42   *The ſecond Booke.*   CHAP.3

Now, there followes the *Sphæricall* triangles, which are moſt hard, as they are commonly deliuered by others; but by our Logarithmes they are the moſt eaſie of all.

Of Sphæricall Triangles
CHAP. III.

Sentences

*IN Sphæricall triangles, the
angle that is neereſt in quantitie to a quadrant, and the
ſides ſubtending the ſame, are
doubtful whether they be of
the ſame, or of a diuers kind,
except the account, or poſition bewray the ſame.*

2   *But euery one of the two oblique angles, is of the
ſame kinde with the ſides ſubtending the ſame.
Therefore knowing of what kinde the one is, it appeareth alſo of what kinde the other is.*

3   *If any angle of a triangle bee neerer to a quadrant then the ſide ſubtending the ſame : two ſides
thereof ſhall be of one kinde, and the third leſſe
then a quadrant.*

4   *But if any ſide of a triangle be neerer to a quadrant then the angle ſubtended thereby, two angles thereof ſhall bee of the ſame kinde, and the
third greater then a quadrant.*

5   *A Sphæricall triangle, is either quadrantall or
not.*

6   *A quadrantal, is that whoſe ſide or angle is e-
quall to a quadrant : whereby we teach, that the
knowledge of a quadrantal that is not right angled
may as eaſily be gotten, as if it were right angled.*

7   *A quandrantall triangle, is either manifold, or
ſingle.*

8   *A manifold quadrantall, is either three right
angled*

CHAP.4.   *The ſecond Booke.*   43

angled, or two right angled.

A three right angled triangle, is that, whereof e-   9
uery part is equall to a quadrant.

Therefore euery triangle, each of whoſe three   10
*parts not being oppoſite, are equall to a quadrant,
is three right angled.*

A two right angled triangle, is that wherof two   11
*angles onely, and the ſides ſubtending them, are ſe-
uerally equall to a quadrant.*

In euery two right angled triangle, the oblique   12
angle, is equall to his ſubtending ſide.

Euery triangle, whereof any part is equall to a   13
*quadrant, and any oblique angle, equall to his ſub-
tendant, is two right angled.*

Euery triangle hauing any two parts ſeuerally   14
*equall to a quadrant, and the third vnequal, is two
right angled.*

All the reſt are called ſingle quadrantals.   15

Of Single Quadrantals.
CHAP. IV.

Single Quadrantall, is that   1
*whereof one part onely is equal
to a quadrant , and the other
fiue parts are not quadrants.*

Of theſe fiue parts which   2
*are not quadrants, thoſe three
which are furtheſt remoued
from the right angle, or the ſide that is a quadrant,
we turne into their complements, and retaining
the old order, we bring them all fiue into a circular, or quinquangled ſituation, and wee call them
circulars.*

Firſt let the triangle BPS be right angled in
B, the fiue oblique parts therof which are not
quadrāts are theſe, BP one of the ſides about
the

the right angle : P
one of the oblique
angles : PS the
fide fubtending
the right angle : S
the other oblique
angle : SB the o-
ther fide about
the right angle,
for which we (for
the eafier calcula-
tion) take the fide

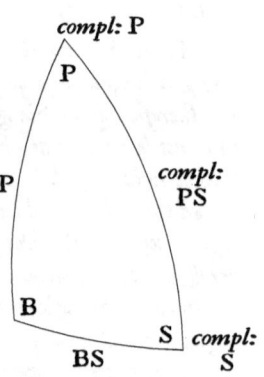

it felfe BP : the complement of the angle P :
the complement of the fide PS : the complement of the angle S : and the fide it felfe
SB, and keeping their naturall fituation, we
place thefe fiue parts in order, as in the margine, and we call them Circulars.

Likewife fecondly, let SPZ be a quadrantall fingle triangle, not right angled (made of
the centers of the Sunne-
rifing, the pole and the
zenith) quadrantall in
the fide ZS, his fiue parts
not quadrants are thefe,
Z one of the angles com-
paffed by the quadrant
fide : PZ the diftance of
the pole from the zenith :
P the angle fubtended by
the quadrant : the fide PS
the diftance of the pole
from the Sunne: And laft-
ly S one of the angles, a-

bout which the quadrant is : inftead wherof
we for our eafier account do take the angle Z
or PZS, being adioyned to a quadrant, and
is

CHAP.4. *The second Booke.* 45

is the arch of the Sunnes diſtance from the North. The complement of PZ, which is the eleuation of the pole : the complement of the angle P, or of the angle ZPZ which is the difference aſcentionall (that is) the difference of the time of the Sunnes riſing or ſetting from ſixe a clocke.

The complement of the ſide PS, which is the declination of the Sunne : and the angle it ſelfe S, or PSZ, which wee call the angle of the Sunnes poſition (that is in reſpect of the Pole and Zenith.) Theſe fiue parts we place in a circular or pentagonall ſite, as in the margine, and we call them Circulars, neither ſhall there be made any other circular parts of the former right angled triangle BPS, if you make P the Pole: S the Sunne: B the Northpoint: for the ſide BP will be the eleuation of the pole: the cõplement of P the diffrence aſcenſionall : the complement of PS, the declination of the Sun : the complement of S the angle of the Sunnes poſition : and laſtly, BS the Azimuth of the Sunne, which are altogether the ſame circular parts that were before, and placed in the ſame ſite towards the left hand that the other was towards the right.

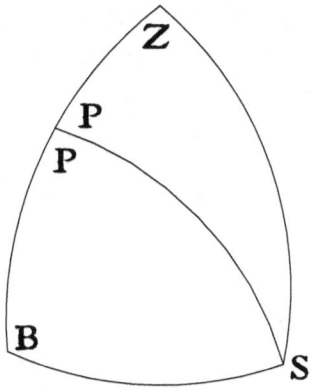

And

46   *The second Booke.*   CHAP.4

And ſo in all quadrantals, as well right-angled, as not.

Corolarie.   3   *Hence it is that there bee many triangles, not conformable in their naturall parts, which in theſe Circular parts, doe altogether agree, and are reſolued by this our methode of Circulars.*

As it clearely appeareth in the two former triangles BPS, and PZS ioyned together : In which all the naturall parts (beſides PS and BS of the former, and PS, and PZS of the latter, do altogether differ, but all the Circular parts agree, as is aforeſaid.

4   *This vniformitie of the Circular parts, moſt manifeſtly appeareth in right-angled triangles made on the ſuperficies of a globe, of fiue great circles, the firſt whereof cutteth the ſecond, the ſecond the third, the third the fourth, the fourth the fifth : and laſtly, the fifth the firſt, at right angles. But all the other Sections ſhall bee made at oblique angles.*

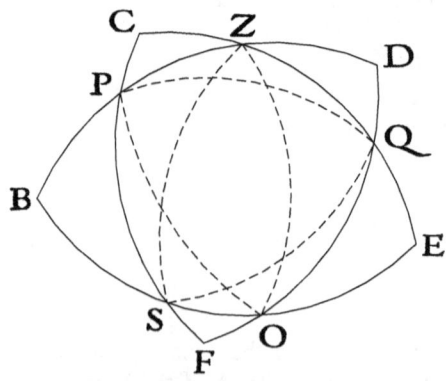

For example ſake : the meridian of any region DB cuts the Horizon BE in the poynt B. The Horizon BE cuts the circle EC, which compaſſeth about the Sunne (that is to ſay, which

CHAP.4. *The second Booke.* 47

which is drawne about the Sun as his pole) in
the poynt E. The circle EC, which compaf-
feth the Sunne, cuts the meridian of the Sun
CF in the poynt C. The meridian of the Sun
CF cuts the Equator FD in the poynt F.
And laftly, the Equator FD cuts the meridian (of that region DB) in the poynt D.
And all thefe fiue Sections in the poynts
B, E, C, F, D, are made orthogonally and at
right angles. The other Sections in the points
Z. P. S. O. Q being made at oblique angles.
There will alfo bee made of thefe Sections
fiue right-angled triangles, PBS, SFO,
OEQ, QDZ, and ZCP, the naturall parts
whereof, although they differ, and are varied
in each feuerall triangle, yet the fiue Circular
parts are the fame that were before without
any difference.

The fame vniformitie of the circular parts ap- 5
peareth alfo in quadrantals that bee not right an-
gled, made vpon the fuperficies of a Globe out of
fiue poynts, the firft whereof is diftant from the fe-
cond, the fecond from the third, the third from the
fourth, the fourth from the fifth, and the fifth from
the firft by diftances and arches equall to a qua-
drant ; but the other diftances of the poynts bee
vnequall to a quadrant. As in the fame figure
the poynts P from Q, Q from S, S from Z,
Z from O, and O from P, are diftant by fpa-
ces equall to a quadrant. But P from Z, Z
from Q, Q from O, O from S, and S from P,
are diftant each from other by arches which
are not quadrants. Thefe will alfo bee made
out of thefe diftances, fiue quadrantals not
right angled, PZQ, ZQO, QOS, OSP,
and SPZ, whereof although the naturall
parts differ, yet the circular parts remaine the
fame

same vnchangeable here as before, that is to say : BP the eleuation of the pole : the complement of BPS, or SPZ, the difference ascentionall : the complement of PS, which is SF, the declination of the Sunne : the complement of PSB, which is PSZ, the angle of the Sunnes position : BS the Azimuth of the Sunne, which doe indifferently agree to all the former triangles, and not to those alone, but also to all triangles which do arise of the other intersections of these ten arches drawn forth, to whole circles, which becaufe they are many, and confused, we here let them passe, it is sufficient to haue warned by this abridgement, that all the confusion of the naturall parts, and of their rules is auoyded, and taken away by these few Circular parts, and their onely rule.

6   *Of the fiue circular parts three alwayes come in question : whereof the two first are giuen, the third is sought for.*

7   *And of these three, one is in the middle, and two are the extreames which are either set about the middle, or opposite to it.*

For example sake. Let the three parts proposed in the question be these : the Azimuth of the Sunne BS : the eleuation of the Pole BP : and the ascensionall difference the complement of BPS, whereof the eleuation of the pole is called the middle, and the other two extreames are called neighbours vnto it, or set about it. But if the three parts coming in question were : the declination of the Sun. the complement of PS : the eluation of the Pole BP : and the angle of the Sunnes position PSZ, the eleuation of the pole shal be called the middle one, as before, but the declination

CHAP.4. *The ſecond Booke.* 49

clination of the Sunne, and the angle of the Sunnes poſition, ſhall bee called the extreames, remoued from the middle, or oppoſite to it. The like reaſon is in the other fiue.

The Logarithme of the middle one is equall **8** *to the Differentials of the extremes ſet about it, or to the Antilogarithmes of the oppoſite extreames.*

This Theorem is proued by induction of all the three parts or triplicities which can be made, and come into queſtion of the fiue circular parts of the former right-angled quadrantall **BPS.** But wee omit the triplicities of the latter triangle not right-angled **PZS.** becauſe all the circular parts thereof are altogether the ſame in quantitie which were in the former (by the 3. 4 and 5. of this chapter.) Now therefore of the fiue circular parts of the right-angled triangle **BPS,** (which are **BS,** or the Azimuth of the Sunne riſing; the complement of **BSP,** or the angle of the Sunnes poſition : the complement of **SP,** or the declination of the Sunne : the complement of **SPB,** or the difference aſcenſinall : and **PB,** or the eleuation of the pole) The 3 which come in queſtion of extremes, ſet about the middle one, are either firſt **BS,** the complement of **BSP,** and the complement of **SP** : or ſecondly the complement of **BSP,** the complement of **SP,** and the complement of **SPB** : or thirdly, the complement of **SP,** the complement of **SPB,** and **PB** : or fourthly, the complement of **SPB,** **PB,** and **BS** : or fiftly, **PB, BS,** and the complement of **BSP.**

But becauſe in all theſe triplicities, *the Tangent of one of the extreames is to the right ſine of the middle one, as the whole ſine is to Tangent of*

D   *the*

the other extreame (as it is manifeſt out of the common demonſtrations of *Trigonometrie*) therefore by our demonſtrations of the *5. Prop.* of the *2.Chap. 1.Book. the Logarithmes of the middle ones* (which are the Logarithme of the middle one onely, by the Corollarie of the ſixt definition of the firſt cha. 1.Book) *are equall to the Tangents of both the extremes.* But *the Logarithmes of the Tangents of theſe extremes are the Differentials of the ſame* (by the 22 and 25 *Sect. Chap. 3. Booke 1.*) Therefore *the Logarithme of the middle one only, is equall to the Differentials of the extremes ſet about it,* as we ſaid in the former part of the *Theorem.* The confirmation of the latter part followeth.

Therefore of the ſame fiue circicular parts the three which come into queſtion of the extreames oppoſite to the middle one, are either, firſt PB the complement of BSP, and the complement of SPB : or ſecondly, BS the complement of SP and PB : or thirdly, the complement of BSP, the complement of SPB and BS : or fourthly, the complement of SP, PB, and the complement of BSP : or fiftly, and laſtly the complement of SPB, BS and the complement of SP.

But in all theſe triplicities, or fiue caſes, *the right ſine of the complement of one of the extreames is to the right ſine of the middle one, as the whole ſine is to the right ſine of the complement of the other extreame* (which is more largely demonſtrated by *Regiomontanus, Copernicus, Lansbergius, Pitiſcus,* and others : then that it can be repeated in this abridgement) therfore by our demonſtrations (the *5.Prop. 2.Cha. 1.Book the Logarithmes of the complements of theſe extreames, are equall to the Logarithmes*

of

the middle ones, that is (as is aforefaid) to the Logarithme of the middle one onely.

But *the Logarithmes of the complements of thefe oppofite extreames, are the Antilogarithmes of the very fame parts,* out of the definition. *Sect.13.* and *16. Chap.3. Book 1.* Therfore it followeth in thefe cafes, that *the Logarithme of the middle one only is equal to the Antilogarithmes of his oppofite extremes,* as the latter part of the Theorē affirmeth. Therfore *the whole Theorem is manifeft.* Befide this proofe now made by induction of all the cafes which can happen, the fame Theorem may bee alfo clearely perceiued by the 4 and 5 of this chapter, in the figure whereof, the like conftitution of the circular parts doth argue the fimilitude of the analogie of the fame. So that whatfoeuer may be truly faid of any middle one, and his extreames fet about, or oppofite, the fame cannot iuftly be denied of the other foure middle ones and their extreames refpectiuely fet about, or oppofed.

A generall Confequent.

HEnce it followeth in *fingle Quadrantals,* 9 *that* out of any two parts giuen, any third fhall be found. *For alwayes* either the middle one is fought for, & his Logarithme is found by adding the Differentials of the giuen extreames fet about, *or* one of the extremes is fought for & his Differential arifeth out of the fubtraction of the Differentiall of the other extreme giuen out of the Logarithme of the middle one already knowne, as in the fiue former triplicities of a right-angled triangle of the Theorem going before, and as many of a not-right-angled triangle : *or elfe* the middle

D 2

one

52   *The second Booke.*   CHAP.4

one is fought for, and his Logarithme commeth forth by adding the Antilogarithmes
of the oppofite extreames giuen. *Or laftly,*
one of the oppofite extreames is fought for,
and his Antilogarithme is found by fubtraction of the Antilogarithme of the other oppofite extreme already giuen out of the Logarithme of the middle one already knowne.
*As in the fiue latter cafes of the right angled trian-
gle of the Theorem going before, and as many of a
not-right-angled triangle. But to euery one of
thefe Logarithmes, Antilogarithmes and Differentials already found, there anfwere two arches
of diuers kindes. Therefore out of the kinde of the
arch fought for, knowne by the 2. 3. or 4. of this*
chapter, or elfe by pofition, *the true arch it
felfe fhall be made knowne.*

As in the former example of the feuenth
Section of this Chapter, three parts of the
queftion are Circular, *The Azimuth of the
Sunne, the Eleuation of the Pole, the Difference
Afcenfionall* : that is, in the right-angled triangle BPS, the parts are BS and PB, and the
complement of SPB : or elfe in the not-right-
angled triangle quadrantall PZS, the parts
are PZS, the complement of PZ, and the
complement of SPZ, of which three let the
extreames fet about be giuen, that is, *The A-
zimuth of the Sunne rifing* BS, *or* PZS, *70 degrees, and the difference Afcenfionall the complement of* SPB, *or the complement of* SPZ, 16
deg. 24 27, and the middle part PB be fought,
or the complement of PZ, which is *the Eleuation of the Pole.*

Let the differentiall therefore of the com-
plement of SPZ 16 degr. 24. 27. $+$ 1222618
Bee added to the Differentiall of BS, or
BZS

CHAP.4. *The second Booke.* 53

B Z S 70 degr. ———————————— 1010683

And there will come forth ──── + 211935 the
Logarithme of BP 54 deg. for the eleuation
of the pole fought for.

An Admonition.

BEfides the *Eleuation of the Pole* thus now
found, there is alfo found by the fame maner of working.

2 *The Azimuth of the Sunne* by the eleuation
of the pole, and the angle of the Sunnes pofition giuen.

3 *The Angle of the Suns Pofition* out of the A-
zimuth of the Sun, and his declination giuen.

4 *The declination of the Sunne* out of the angle of the Sunnes pofition, and the difference
Afcenfionall giuen.

5. *The Difference Afcenfionall* out of the Dedi-
nation of the Sunne, and Eleuation of the
pole giuen.

The fecond Example.

LEt the *Azimuth of the Sun rifing* bee giuen
BS, or PZS 70 degr. and the *Eleuation of
the Pole*, 54. *degr.* which is PB, or the comple-
plement of PZ : and let the *Difference Afcen-
fional* be fought, that is, the cöplement of SP
B, or the complement of SPZ. And becaufe
here likewife the extreame parts are fet a-
bout the middle part, therefore

Take the Differentiall of the Suns Azimuth,
BS, or BZS 70 deg. which is ——— 1010683
Out of the Log. of the eleua-⎫ + 211935
tion of the Pole BP, 54. deg ⎬ ———————
And there will come forth ——+ 1222618 the
Differentiall of SPB 16 deg. 24′ 27″ the arch
of the *Afcenfionall difference* fought for.

D 3   *An*

54   *The second Booke.*   Chap.4

An Admonition.

IN imitation of this example there is found
2 *The Declination of the Sunne* out of the difference Afcenfionall, and the Eleuation of
the pole giuen.
3 *The Angle of the Sunnes pofition* out of the declination of the Sunne, and Difference Afcenfionall giuen.
4 The *Azimuth of the Sunne* out of the angle
of the Sunnes pofition, and the declination
of the fame giuen.
5 *The Eleuation of the Pole* is had out of the
Sunnes Azimuth, and the angle of the Suns
pofition giuen.
    Alfo contrariwife there is found
6 *The Difference Afcenfionall* out of the declination of the Sunne, and the angle of the
Sunnes pofition giuen.
7 *The declination of the Sunne* out of the angle
of the Sunnes pofition, and his Azimuth giuen.
8 *The Angle of the Sunnes Pofition* is had out of
the Azimuth of the Sunne, and the Eleuation
of the pole giuen.
9 *The Azimuth of the Sunne* out of the Eleuation of the pole, and the Difference Afcenfionall giuen.
10 And laftly, *the eleuation of the Pole* is found
out of the difference Afcenfionall, and the
Declination of the Sunne giuen.

The third Example.

IN the latter example of the fame 7ᵗʰ Section of this chapter, thefe three circular
parts of the queftion are propounded, *The
Declination of the Sunne, the Eleuation of the
Pole*

CHAP.4.   *The second Booke.*   55

Pole, and *the Angle of the Sunnes poſition.* Theſe in the right angled triangle BPS, are the complements of PS, BP, and the complement of BSP. And in a not-right-angled quadrantall PZS, they are the complement of PS, ZP, and ZSP. Of which three let the oppoſite extreames be giuen, that is, *the Declination of the Sun,* which is the complement of PS 11 deg. 35. 51. and *the angle of the Sunnes poſition,* which is the complement of BSP, or ZSP 34 degr. 19. 21. almoſt. And let the middle part BP be ſought, or the complement of ZP, *which is the eleuation of the Pole.* Therefore

Let the Antilog. of the complement of PS 11 degr. 35. 51. which is —+ 20627

be added to the Antilog ⎱ + 191308 and
of BS, 34. degr. 19. 21. ⎰ ————————

there will come forth—— + 211935 the Logarithme of BP 54 degr. for the *Eleuation of the pole* that was ſought for.

An Admonition.

BEſides the Eleuation of the pole now firſt found after this manner, you may by the ſame practiſe haue

2 *The Azimuth of the Sunne* out his declination and the difference Aſcenſionall giuen.

3 *The angle of the Sunnes poſition* out of the difference Aſcenſionall, and the Eleuation of the the pole giuen.

4 *The Declination of the Sunne* out of the Eleuation of the pole, and Azimuth of the Sun giuen.

5 And laſtly, you ſhall finde *the Difference Aſcentional* out of the Azimuth of the Sunne, and the angle of the Sunnes poſition giuen.

D 4   *The*

56   *The fecond Booke.*   CHAP.4

The fourth Example.

L Et *the Declination of the Sunne* bee giuen the complement of SP 11 degr. 35. 51. And *the Eleuation of the pole* BP, or the complement of PZ 54 degrees. And let *the angle of the Sunnes pofition* the complement of BSP, or PSZ be fought for. And here likewife, becaufe the extreame parts are oppofed to the middle, therefore

Out of the Log.of B P 54 deg. $+$ 211935
take the Antilog. of the complement of PS 11. deg. 35. 51. $+$ 20627

and there will remaine ————$+$ 191308 the Antilogarithme ofthe complement of BSP. 34. deg. 19 21. almoft, the angle of the *Pofition of the Sunne* fought for.

An Admonition.

B Efides the angle of the Sunnes pofition found out by this firft practife, there is found

2 By the fame practife *the Declination of the Sunne,* out of the difference Afcenfionall, and the Azimuth of the Sunne giuen.

3 The *Difference Afcenfionall* is found out of the Eleuation of the pole, and angle of the Sunnes pofition giuen.

4 *The Eleuation of the pole* is found by the Sunnes Azimuth, and his declination giuen.

5 *The Azimuth of the Sunne* is found, out of the angle of the Sunnes pofition, and the difference Afcenfionall.

6 In a contrary order, *The Angle of the Suns Pofition* is found by the Azimuth of the Sun, and the difference Afcenfionall giuen.

The

CHAP.5. *The second Booke.* 57

7 The declination of the Sunne is had out of the angle of the Sunnes pofition, and the eleuation of the pole giuen.

8 The Difference Afcenfionall is found out of the Sunnes declination, and Azimuth giuen.

9 The Eleuation of the Pole is had out of the Difference Afcenfionall, and the angle of the Suns pofition being giuen.

10 And laftly, *The Azimuth of the Sunne* is found by the Eleuation of the pole, and the Suns declination giuen.

And fo in imitation of thefe foure examples, thirtie feuerall queftions of Circular parts in a right-angled quadrantall, and as many in a not-right-angled quadrantall, are refolued by this generall Confequent, by the benefit of one Addition or Subtraction onely. But for the vnderftanding of the latter part of this Confequent, of the kindes of arches, fee the 3. 4. 5 and *6* Examples of the Chapter following.

Of Not-quadrantals mixt.
CHAP.V.

Itherto hath beene taught the doctrine of quadrantall Sphæricall triangles : there followeth now the doctrine of Sphæricall triangles not quadrantall.

1 A not quadrantall is a Sphæricall triangle, whereof neither fide nor angle is a quadrant.

2 A not-quadrantall is reduced to two quadrantals, if from the top either a perpendicular or a

D5 *quadrant*

58   *The second Booke.*   CHAP.5

quadrant arch be let downe to the bafe (extended
as need fhal be)

3   The perpendicular falles within the triangle, if the angles at the bafe bee both of one kinde ;
but it falles without if they bee of diuers kindes,
and contrariwife.

4   The quadrant arch falles without the Triangle
if the legs be of one kinde ; but within if they bee
of diuers kindes, and contrarife.

5   Out of the fixe parts of a not quadrantall three
giuen only, are fufficient to get knowledge of the
reft, except of the three giuen, whereof one is oppo-
fit to the other, the third be neerer to a quadrant,
then the other giuen of the fame kinde : for in this
cafe it is required alfo, that the kind of the part
which is oppofite to the third be alfo giuen, that the
other parts may be knowne.

Examples of this cafe are the 4 and
6 examples following.

6   The three parts giuen are either mixt or pure.

7   They are mixt whereof one is of a diuers kinde
from the other two : As when two fides and any
angle are giuen, or two angles with any fide.

8   In mingled parts giuen, if from that tearme of
the fide giuen, in whofe other terme is the angle giuen, a perpendicular or a quadrant arch fubtending that angle, fall to the bafe, the not-quadrantall triangle fhall be reduced to two quadrantals that may be known by the 9 Sect. of the 4 chap
of this booke.

And therefore, the parts of a not-quadrantall,
becaufe they are all one with the parts, or remainders of thefe parts to a femicircle, are eafily known,
the kindes of the parts being yet firft knowne by
the fecond, third, and fourth Section of the third
chapter of this booke, or elfe by pofition.

An

CHAP.5.   *The second Booke.*   59

An Example of two sides, and the angle
betweene them giuen.

A S for vse and exercise sake, let there be a
sphærical triangle not quadrantal descri-
bed on the superficies of the *Primum Mobi-*
le PZS representing the pole, the zenith, and
the Sunne ; whereof there be six parts.

The side PZ, which is the distance of the
pole from the
zenith,   or
the comple-
ment of the
poles eleua-
tion.

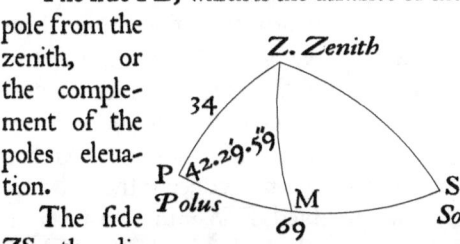

The side
ZS, the di-
stance of the zenith and Sunne, or the complement of the Sunnes height.

The side PS, the distance of the pole and
the Sun, or the complement of the declination of the Sunne from the Equator.

The angle ZPS, the houre of the day, or
the degrees of the Equator.

The angle PZS, which is the Azimuth of
the Sunne frome the North.

The angle PSZ, which is the angle of the si-
tuation and position of the Sunne to the pole
and zenith.

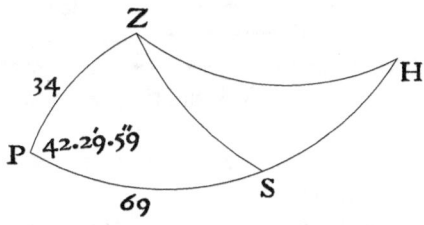

Of these six parts let any three be giuen,
partly

60   *The second Booke.*   CHAP.5

partly angles, partly fides. *For example fake.*

The houre angle ZPS 43 degr. 29ʹ 59ʺ which fheweth two of the clock afternoone 49ʹ 59ʺ. and 56ʺ.

And the fide PZ 34. *the complement of the eleuation of the Pole.*

And the fide PS 69 *the complement of the declination of the Sunne.*

Out of which that the other three partes may bee gotten : from Z the end of the fide PZ that was giuen, let the perpendicular ZM, or rather (if you will) the quadrant ZH be drawne downe, fubtending the angle ZPS, and reducing the not-quadrantall propofed PZS into two triangles quadrantall in the angle M, which are PMZ, ZMS, as in the firft figure : or if you bee delighted with varietie, let them be reduced to two triangles quadrantall in the fide ZH, which are ZHP, and ZHS, as in the fecond figure. All the parts of which quadrantals you fhall get by the ninth Section of the 4ᵗʰ Chapter of this booke.

For by hauing PZ giuen ——— 34 deg.
and ZPM, or ZPS ——— 42 deg.29ʹ59ʺ
You may find the perpend. ZM 22 deg.11.47
And the angle PZM——— 52 deg.46.38
And the fide PM ——— 26 deg.26.29
Which PM being taken out ⎱ 69 deg.
      of PS ——— ⎰ ———————
There remaines MS ——— 43 deg.33.31ʺ.

Now the fide MS, and the perpendicular ZM being knowne, you may (by the faid ninth Section of the fourth chapter of this booke,) finde out

The angle oppofite to the perpendicular MSZ, or that which was fought for PSZ 31.deg.6.5ʺ.

CHAP.5.   *The second Booke.*   61

And the fide that was fought SZ. 47 deg.
And the angle MZS————— 67 de.38. 11
Which being added to PZM— 52 de.46.38

Makes PZS the angle fought   120 de.24.49
    You haue therefore three parts which you
fought for, found by helpe of the perpendicular ZM of the former figure.
    You may alfo finde the fame by helpe of
the quadrant ZH in the latter figure.
For hauing PZ giuen————34 deg.
And ZPS, or ZPH ————32 deg.29.59.
You may finde by the fame 9. Sect. of the 4.
chap. of this booke,
        The angle ZHP———— 22 deg. 11.47.
    And the angle PZH——— 142 deg. 46 38.
    And the fide PH———— 116 deg 26. 29.
Out of wth PH fubtract PS 69 deg.
there remaines the fide SH   47 deg. 26. 29.
    Which fide SH being now had, together
    with the angle ZHP   22 deg .11 47. you
may alfo (by the faid 9 Section of the 4 chap.
of this book) finde out
The angle HSZ ———— 148 deg.53.55.
And the remainder⎤
therof to a femicir- ⎬——— 31 deg 6.5 that
cle, the angle PSZ ⎦   was fought for.
And the fide SZ ———— 17 deg. that was
fought for.
And laftly the angle HZS   22 deg. 21.49.
which being taken out⎤   142 deg. 46.38.
     of HZP ——⎦   —————
There remaines PZS — 120 deg 34 49. the
other angle that was fought for, in all poynts,
as before.

                An Admonition.

I n imitation of this example, nine diuerfe
                                        queftions

queſtions may be reſolued both of this, and any triangle. For by *the Eleuation of the Pole,* and *the houre of the day,* and *the declination of the Sunne that day,* being giuen, there is had, as afore :

1 The Azimuth of the Sunne.

2 The height of the Sunne.

3 The angle of poſition of the Sunne : alſo by ha- uing *the declination of the Sunne, the angle of the Sunnes poſition, and his height* giuen, you haue

4 The Sunnes Azimuth.

5 The Eleuation of the Pole,

6 The houre, or houre arch.

Alſo if you haue *the height of the Sunne, his A- zimuth,* and *the height of the pole* giuen, there is found,

7 The houre of the day.

8 The declination of the Sunne.

9 And *the angle of the Sunnes Poſition.*

The ſecond example of two angles giuen, and the ſide betweene them.

THe angles in the figures going afore, be- ing giuen, to wit,

    The houre angle   ZPS 42 deg.29 59″.

& *the azimuth of the ſun* PZS. 120 deg.24 49″.

with the ſide between them, being *the compl.* $\Big\}$ PZ 34 deg.

of the poles eleuation

The other 3 parts are ſought out. For as a- fore,

    Hauing firſt giuen   ZM. 22. deg. 11 47″.

    And ――――― PM. 26 deg. 26 29″.

    And the angle― P Z M. 52 deg. 46. 38″.

wᵗʰ being taken out of PZS. 120 deg. 24 49″.

there being left remai. MZS. 67 deg. 38. 11″.

                                 By

CHAP.5.   *The second Booke.*   63

By which MZS, and ZM, already known,
there shall at length be found,

The side ZS —— 47 deg. the side
sought for.

And the angle ZSM, or ZSP 31 deg 6.5.
the angle sought for.

And the side MS.   42.deg.33.31.
which being added to PM.   26deg.26.29.

the side remaineth—— PS 69 deg which was
sought for.

And these you haue by meanes of the perpendicular of the former figure. In like manner you may finde the same by helpe of the
quadrant of the latter figure. For they are
found by the ninth Section of the fourth
chapter of this booke.

by hauing giuen the angle PHZ. 22,de.11.47

And the angle —— PZH.142.de.46.38

Out of which the angle⎫   120 de.24.49
giuen PZS being giuē ⎬   ————————
There remaines —— SZH. 22 de.21.49

which together with the angle PHZ. now
knowne, all the rest of the parts are brought
forth. viz.

PZ.	34 deg.
ZPS.	42 deg.29. 59.
PS.	69 deg.
PS Z.	31 deg. 6. 5.
SZ.	47 deg.
ZSH.	148 deg.53.55.
SH	47 deg.26.29.

An Admonition.

IN imitation of this example, nine diuerse
queftions of this and of any other triangle,
are refolued.

For *the houre of the day, the Eleuation of the
pole,*

64   *The ſecond Booke.*   Chap.5

pole, and *the ſuns azimuth* being giuen, there is had,

    1 The declination of the ſunne,
    2 The angle of the ſunnes poſition.
    3 The height of the ſunne.

Alſo, *the houre of the day, the declination of the ſunne,* and *angle of the ſunnes poſition* being giuen, there is had,

    4 The height of the ſunne,
    5 The ſunnes azimuth,
    6 The height of the pole.

Alſo, *the angle of the ſunnes poſition, the height of the ſunne,* & his *azimuth* being giuen, there is had,

    7 The height of the pole,
    8 The houre of the day,
    9 The declination of the ſunne.

The third example of two ſides giuen, whereof that which is neereſt to a quadrant, ſubtendeth the angle giuen.

I N the figures afore, let there be giuen
            The ſide— PZ. 34 deg.
And that which is neerer ⎫
then it to a quadrant, —⎬ ZS. 47 deg.
With that angle which ⎫
this ſide ſubtendeth—⎬ ZPS. 42 de. 29.59
By the 9 Sect. of the 4.chap. of this Booke, let there be ſought out
        The ſide ZM 22 deg. 11 47
      And the angle PZM 52 deg. 46 38.
       And the ſide PM 26 deg. 26 29.
    And in like manner you may haue
            ZSM 31 deg. 6.5. or ZSP
the angle ſought for : which is moſt certain-
ly known (by the 2 Sent. 3.chap. of this book)
                           to

CHAP.5.   *The fecond Booke.*   65

to be leffe then a quadrant, namely, that it
is 31 deg $6.\ddot{5}$. ad that it is not 148 deg.$5\ddot{3}.\ddot{5}\ddot{5}$

Alfo you may haue⎫
the angle ——⎭ MZS 67 deg.38.$\ddot{11}$

which being added to   PZM52 deg.4$\ddot{6}.\ddot{3}8$

makes the other angle —— PZS 120 deg.24.49
which was fought for.

Laftly, you fhall haue⎫ MS 42 deg. $\ddot{3}\ddot{3}$. $\ddot{3}\ddot{1}$
the eafe —— ⎭

Which being added to MP 26 deg.16.29

makes the fide —— PS 69 deg. which
was fought for.

No otherwife (if you will) you may find out
the fame by helpe of the quadrant ZH of the
latter figure.

The fourth example, of two giuen fides, whereof
that which is leffe neere a quadrant fubten-
deth the angle giuen, and that which is
neereft, fubtendeth an angle of the
kinde giuen onely.

A dmit there begiuē⎫ Z S 47 deg.
the fides ——⎭

And that which is not ⎫ PZ 34 deg.
fo neere a quadrant ⎭

with that angle which ⎫ ZSP 31 deg. $6.\ddot{5}$
this fubtendeth ⎭

And let it be giuē, that the angle wch ZS fubtendeth, that is the angle SPZ, is by kind leffe
then a quadrant : therfore the perpendicular
ZM being let down from Z to the bafe PS (as
before) or the quadrant ZI(as here) fubtending the giuen angle ZSP. By the 9 Sect. of
the 4 chap of this booke, let the other parts
be gotten (as for exercife and varieties fake)
by the quadrant of this figure, ZI, you may
get

The

66   *The fecond Booke.*   CHAP.5

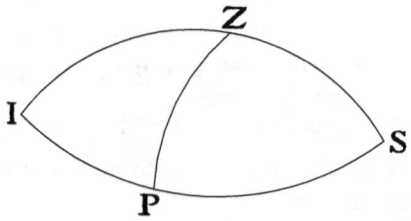

The angle ZIS ——— 22 deg. 11. 47.
And —— IZS ——— 157 deg.38.11.
And——— SI ——— 132 deg.33.31.
and in like maner ⎱
you may haue ⎰ IPZ ——— 137 deg.30.-1.
& by confequent ⎱
the angle —— ⎰ SPZ ——— 42 deg.29 59.
that was fought for.

Becaufe by the giuen pofition, it is expref-
ly declared to be leffe then a quadrant : o-
therwife except the kinde therof were giuen,
it would be vncertaine (by the 1.Sect. of the
3.chap. and the 5 Sect. of this chap.) for it
might otherwife haue been 137 deg.30.1.
So alfo fhall you haue ⎱
the angle ⎰ IZP. 37 deg.13.22
which being taken out of IZS. 157 deg.38.11
there remaines the ⎱
other angle fought. ⎰ PZS. 120 deg.24.49
To conclude, you ⎱
fhall alfo haue ⎰ —IP. 63 deg.33 31
Which being taken ⎱
out of ⎰ — IS. 132 deg.33.31
There remaines ——— PS. 69 deg. the
fide that was fought for.

You fhall alfo hit the fame markes (as it
were) if you feeke the account or number
of the parts, by helpe of the perpendicular
ZM of the firft figure.

An

CHAP. 5.   *The second Booke.*   67

An Admonition.

By imitation of the third example going before, and this fourth, there are refol-ued eighteene diuerfe queftions of this and any other triangle. For (as it is in the 3 example) *the height of the pole, the height of the Sun, & houre of the day,* being giuen, there is found,

1 *The Azimuth of the Sunne.*
2 *The angle of the Sunnes pofition.*
3 *The declination of the funne.*

Alfo, the *Eleuation of the pole, the height of the funne,* and *the angle of the funnes pofition* being giuen, (as in the fourth example) there are found,

4 *The Azimuth of the funne,*
5 *The houre of the day,*
6 *The declination of the funne.*

Alfo, *the height of the funne, the declination of the funne,* and *houre of the day being giuen,* there is found,

7 *The angle of the funnes pofition,*
8 *The Azimuth of the funne,*
9 *The Eleuation of the pole.*

Alfo, *the height of the funne, the declination of the funne,* and *Azimuth of the funne* being giuen there is found,

10 *The angle of the funnes pofition,*
11 *The houre of the day,*
12 *The Eleuation of the pole.*

Alfo, *the declination of the funne, the Eleuation of the pole,* and *the angle of the funnes pofition* being giuen, there is found,

13 *The Azimuth of the funne,*
14 *The height of the funne,*
15 *The houre of the day.*

Alfo, *the declination of the fun, the height of the pole,* & *azimuth of the fun being giuen,* you haue

The

68   The second Booke.   CHAP.5

16 *The houre of the day,*
17 *The angle of the funnes pofition,*
18 *The height of the funne.*

The fifth example of two angles giuen, the neerer
whereof to a quadrant is fubtended
by the fide giuen.

IN the Triangle PZS of the firft Dia-
gram,

    Let the angle PSZ   ⎱ ⎰ —— 31. 6. 5́
    bee giuen ——

And SPZ which is nee-⎫
rer then it to a qua-⎬ —— 42. 29́. 59̋.
drant ————⎭

With the fide ZS fub-⎫ —— 47. 6́. 6̋.
tending the fame ⎭

Out of which PSZ, ⎫
and SZ, is found the ⎬ —— 22. 11́. 47̋.
perpendicular SM—⎭

    (By the 9ᵗʰ of the 4ᵗʰ of this booke.)

And the other parts ⎫
of the quadrantal, ⎬ ——MZS. 67. 38́. 11̋.
SZM, to wit, —— ⎭

    And the fide   —— MS. 42. 33́. 31̋

As alfo by this perpendicular, with the angle ZPS being giuen, or the angle ZPM,
all the partes of the quadrantall ZMP
are found.

As firft the fide fought for PZ. For this
is moft certainely knowne (by the fecond
Sentence of the firft Chapter of this book)
to be leffe then a quadrant, namely, that it
is 34, and not 146.

                            Then

CHAP.5. *The second Booke.* 69

Then we haue PZM. 52, 46, 38,
W^{ch} being added to SZM. 67, 38, 11,
there is made the angle PZS .120, 24, 49,
Laftly, there is alfo had PM. 26, 26, 29,
W^{ch} being added to — MS. 42, 33, 31,

There is made the o-⎤
    ther fide,——⎦ P S. 69, 6. 0, that
was fought for.

You may alfo finde out thefe parts otherwife (if you wil) by the two quadrantals of the
figure next going before, ZIS, and ZIP.

The fixth example of two giuen angles, whereof
that which is not neereft a quadrant is fubtended
by the fide giuen, and that which is neereft, is
fubtended by a fide, whofe kinde
onely is giuen.

OF the triangle PZS of the firft figure,
   let there be giuen,
            The angle, ZPS, 42, 29, 59,
And that which is not fo ⎤
neere to a quadrant as it ⎬ ZSP. 31, 6, 5,
With the fide fubten ⎤
    ding the fame ⎦ —— PZ. 34. 6. 0,
And let it be knowne that the fide ZS, fubtending the angle ZPS, is leffe then a quadrant.

By thefe thus giuen, let ⎤
the perpend: be fought for. ⎦ ZM. 22. 11. 47
And the other parts of the quadrantall
PZM.
        To wit the angle PZM. 52, 46, 38,
        And the fide ——PM. 26, 26, 29,
As alfo by this perpen-⎤
dicular, together with ⎬ZSM or ZSP. 31, 6, 5
the angle giuen —— ⎦
                                Let

Let all the parts of the quadrantal ZMS
be fought,

As firft the defired fide ZS 47. ó. ó. becaufe that by pofition giuen, it is expreffely
declared to bee leffe then a quadrant, otherwife it might haue been here 133. For (by the
firft Chap 3 and 5 of this booke) it is vncertaine, except the kinde thereof be expreffely giuen,

       Then the angle MZS — 67. 38. ıı̋
which added to the angle MZP — 52. 46. 38

       Maketh the angle P Z S. 120. 24. 49
which was defired.

Laftly, there is alfo ⎱
       obtained   ⎰ — SM. 42, 33, 31

Which added to the fide — PM. 26, 26, 29
Maketh the bafe defired — PS. 69, ó, óő

You may alfo moft eafily get the fame parts
out of the two quadrantals PHZ, and SMZ
of the fecond figure.

An Admonition.

BY imitation of the 5 example going afore,
and this fixth, 18 feuerall queftions of
this and any other Triangle are refolued.

For (as in the fifth example) out of the
*Angle of the pofition of the funne, the houre of
the day,* and *height of the funne* being giuen, is
gotten,

1 *The Eleuation of the pole,*
2 *The Azimuth of the funne,*
3 *The declination of the funne,*

    Alfo (as in this fixth example) by *the houre
of the day, the angle of pofition of the funne,* and
height of the pole being giuen, there is gotten,

                             The

CHAP.5. *The second Booke.* 71

4 *The height of the sunne,*
5 *The sunnes Azimuth,*
6 *The declination of the sunne.*

Alſo by *the houre of the day, the azimuth of the sunne,* and *the height of the sunne* giuen, there be gotten,

7 *The declination of the sunne,*
8 *The angle of the sunnes poſition,*
9 *The height of the pole.*

Alſo by *the houre of the day, the Azimuth of the sunne,* and *the sunnns declination* giuen, you haue,

10 *The height of the sunne,*
11 *The angle of the sunnes poſition,*
12 *The height of the pole.*

Alſo by *the Azimuth of the sunne, the angle of the sunnes poſition,* and *the declination of the sun* giuen, there bee gotten,

13 *The height of the pole,*
14 *The houre of the day,*
15 *The height of the sunne.*

Alſo by *the azimuth of the sunne, the angle of poſition of the sunne,* and *height of the pole* giuen, there be gotten,

16 *The declination of the sunne,*
17 *The houre of the day,*
18 *The height of the sunne.*

And ſo by the method of this Canon onely, foure and fiftie ſeuerall queſtions of the ſame triangle, not being quadrantall, are reſolued. The reſt ſhall bee reſolued hereafter.

By theſe therefore it is manifeſt, that of two an- 9
gles, and their ſubtending ſides, three being giuen the Logarithme of the fourth of them at the leaſt, ſhal be made known euen without any deſcription at all of the quadrantals. For out of the ſum of the Logarithmes of the angle and ſide adioyning there-

to

72   *The second Booke.*   CHAP.5

to being giuen, subtract the Logarithme of the
third thing that is giuen, and thence shall come
the Logarithme of the fourth that was sought for;
and that fourth it selfe shall also be made knowne
if the kinde thereof be not vnknowne.

As may be perceiued by the third, fourth,
fifth, and sixth examples going before. For
of the angles of the base, ZPS, and ZSP,
and of their subtending leggs ZS, and ZP.
let three bee giuen, (for examples sake)
The legs ZS. 47 deg. and his Logar. 312858
And — ZP. 34 deg. and his Logar. 581261
with the angle adioy- $\rbrace$ & his Logar. 392172
ning, ZPS. 42, 29, 59

added to the Logar. of ZP last mentioned,
Their summe is, ————————— $+$ 973433
(which is the Logarithme of the secret and
suppressed perpendicular ZM, or of the angle ZHS, or ZIP.)

Out of which subtract $\rbrace$ ———— $+$ 312858
the Logar. of ZS ——

There remaineth the $\rbrace$ ———— $+$660575
Log. of the 4th ZSP.
that was sought for.

Therefore the same fourth it selfe ZSP. will
be 31. 6. 5. Because (by the second Section
of the third chapter) it is proued to bee lesse
then a quadrant.

         Now contrariwise,

There being giuen Z S $\rbrace$ ———— $+$ 581261
34 deg. and his Logar.

And ZS 47 deg. and his Logar. $+$ 312858

wth the angle adioyning $\rbrace$ & his Log. 660575
thereto ZSP, 31, 6. 5.

added to the Logarithme of $\rbrace$ ————
the side last mentioned, ZS,

The summe is ————— $+$ 973433
                     Out

CHAP.*6.*   *The second Booke.*   73

Out of which take the Log. of ZP $+$ 581261

Tere will emaine the   ⎫
                     ⎬ ——— $+$ 392172
Lgarithme of the 4th ⎭

tht was fought for : that is, ZPS, whofe
arc (by the 1 *Sect.* 3. *(hap.*) is vncertaine
whther it be 42, 29$'$, 59$''$, or 137, 30$'$, 1$''$, except
it be knowne by pofition giuen whether it be
greater or leffe then a quadrant.

Of not-quadrantals which be pure.
CHAP. VI.

Itherto wee haue fpoken of
intermingled parts giuen :
now follow fuch parts as
are pure.

      They are pure when the   5
*three parts giuen are of the
fame kinde, and they are either
three fides giuen, and the angles are fought for : or
the three angles giuen, and the fides are fought
for.*

An Admonition.
Although the pure parts are the former in re-   6
*gard of their fimplicitie, yet for their difficultie
they do worthily take the latter place.*

In Sphæricall Triangles.
Halfe the bafe, and halfe the difference of the   3
*legs being taken together, and the Logarithme
thereof, and the Logarithme of the difference of
them being added together; and out of that fumme,
the fumme of the Logarithmes of the legs being
fubducted, the halfe of that which remaineth is
the Logarithme of halfe the verticall angle.*

      Becaufe *Regiomontanus* in the fecond chap-
ter of his fifth booke of Triangles, and others

          E   do

74   *The ſecond Booke.*   CHAP.5

do teach, that *as the rectangle comprehëded vn-*
der the right ſines of the legs, is to th ſquare of the
whole ſine : ſo the difference of the verſed ſine of
the baſe, and difference of the legs is to the ve'ed
ſine of the verticall angle. Seeing alſo as tat
difference is to this verſed ſine : ſo is the rct-
angle made of the right ſines of the ſumne,
and difference of the halfe baſe, and halfe
difference of the leggs, to the ſquare of the
right ſine of halfe the verticall angle (for this
laſt rectangle is to that difference of the verſed ſines, and this laſt ſquare to that verſed
ſine in a 5000000ᶠᵒˡᵈ proportion, the whole
ſine being 10000000) therefore it ſhall follow, that *as the rectangle contained vnder the*
right ſines of the leggs, is to the ſquare of the whole
ſine, ſo ſhall the rectangle made of the right ſines
of the ſumme and difference of the halfe baſe and
halfe difference of the leggs, be to the ſquare of the
right ſine of halfe the verticall angle. And by
conſequent (out of the Corolarie of the ſixt
Definition of the firſt Chapter, & the fourth
Propoſition of the ſecond Chapter, and third
Problem of the fift Chapter firſt Booke) *The*
ſumme of the Logarithmes of the leggs ſubtracted
out of the Logarithmes of the ſumme and diffe-
rence of the halfe baſe, and halfe difference of the
leggs, leaueth the double of the Logarithme of half
the verticall angle, as is aboue ſaid.

4   *Secondly, halfe the baſe, and halfe the aggregate of*
the legs being taken together, and the Logarithme
thereof, and the Logarithme of the difference of
them being added together; & out of that ſumme,
the ſumme of the Logarithmes of the legs being
ſubtracted, the halfe of that which remaineth is
the Antilogarithme of halfe the verticall angle.

For the ſumme of the Logarithmes of the
ſumme

CHAP.*6.* *The second Booke.* 75

fumme and difference of the halfe bafe, and
halfe fumme of the legs of this propofition,
hath no other proportion to the fumme of
the Logarithmes of the fumme and difference of the halfe bafe, and halfe difference
of the legs of the former propofition, then
the double of the Antilogarithme of halfe the
verticall angle here, hath to the double of the
Logarithme of the fame halfe verticall angle
before : The demonftration whereof belongeth to another place.

An Admonition.

I N *Sphæricall triangles alfo, we take the true &*   5
Alterne bafe in the fame fence as before in
right lined triangles, that is, the one for the fum,
and the other for the difference of the cafes.
Thirdly, the Differentiall of the aggregate, and the   6
Differential of half the difference of the legs being
added together, and out of the fumme thereof the
differentiall of halfe the true bafe being fubdu-
Eted, there will remaine the Differentiall of halfe
the Alterne bafe.

The fundamentall reafon hereof is, becaufe that *as the Tangent of the true halfe bafe*
is to the Tangent of halfe the fumme of the legs,
fo is the Tangent of halfe the difference of the legs
to the Tangent of the Alterne halfe bafe. For the
Logarithmes of Tangents are the Differentials of their arches, by the 22 and 25 Sect.
3. Chap. 1.Booke.

And therfore fhall that equalitie of the Logarithmes or Differentials follow this analogie of the Tangents (by the 4 Prop. 2. Chap.
1. Booke.) But becaufe the readers hereof
wil perhaps require of me the demonftration

E 2   of

of this fundamentall analogie, or proportion of Tangents (hitherto vnknowne) I will here therefore fhew the fame, fo farre forth as the fhortneffe of this abridgement will permit.

Let the Sphære therefore AFPG lye vpon the flat fuperficies HIKQ, that they may touch each other in the common poynt A : from which by the center of the Sphære ⊙, let the right line A⊙P be raifed, cutting the vpper halfe of the Sphære in the poynt P, and fo A⊙P fhall bee perpendicular to the plaine or flat HIKQ. Then from the angle A, let be defcribed vpon the fuperficies of the Sphære, the triangle Aλγ fharpe angled in γ, or Aλβ blunt angled in β, and the femicircles AλP, and AγP, or AβP being drawne forth, taking λ for the pole, according to the diftance λγ, or λβ which is equall thereto, draw the circle εδβγ, cutting λP in ε, and λA in δ, and Aβγ in the points β and γ. From the poynt λ to the Arch Aβγ let downe the perpendicular Arch λμ. Here therefore Aλ fhall be the greater leg, and λγ or λβ the leffe leg, Aλ and Aβ the bafes, the one true, the other alterne, Aδ the difference of the legs, and Aε the fumme of the legs, becaufe λε and λδ by the conftruction, are equall to the leffe leg λγ or λβ. This being done, and fuppofing P to be inftead of an eye, or fome lightfome body, from the fame P to the flat lying vnder HIKQ, let downe the beame Pγ, cutting the flat in *c*, and the beame Pβ cutting the flat in *b* : and becaufe λβA are in the fame plaine or circle with the lightfome body P, their fhadows *cb*A fhall be in the fame right line.

Likewife

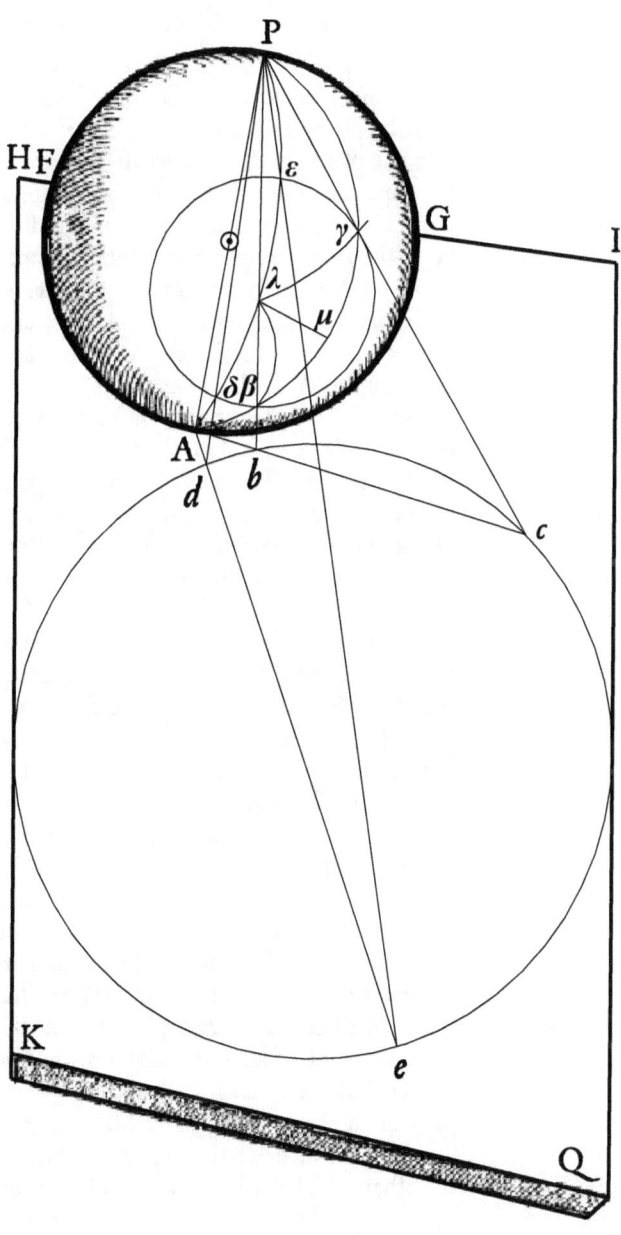

E2

Likewiſe from the ſame poynt P to the ſame plaine, let downe the beame Pε ., cutting the plaine in *c*, and the beame Pδ cutting the plaine in *d*, and becauſe *c*δA are in the ſame plaine and circle with the lightſome body P : therefore their ſhadowes *ed*A ſhall be in the ſame ſtreight line. Moreouer, becauſe P$\odot$A is perpendicular to the plaine, therefore the triangles PA*d* and PA*e*, and PA*b* and PA*c* are right angled in A : and therfore alſo A*d* is the Tangent of the angle APδ or AP*d* & A*e* the tangent of the angle APε or AP*e*. So alſo A*b* is the tangent of the angle APβ, or AP*b* and A*c* is the Tangent of the angle APγ or AP*c* ſuppoſing PA to be the gnomon or whole ſine. And becauſe A*d* is the Tangent of the angle APδ, and APλ is the halfe of the angle A$\odot\delta$ (by the 20 *Prop.* 3. *Eucl.* becauſe this angle is in the center, and that in the circumference) therefore A*d* is the tangent of halfe the angle A$\odot\delta$, or which is the ſame of halfe the Arch Aδ, which is the halfe difference of the legs. Likewiſe becauſe A*e* is the tangent of the angle APε, and the angle APε in the circumference is halfe the angle A$\odot\varepsilon$ in the center, therefore A*e* is the tangent of halfe A$\odot\varepsilon$, or of halfe the arch Aε, which is the halfe ſumme of the legs. In like manner in the baſes true and alterne, A*b* ſhall bee the Tangent of the angle APβ, or of halfe the angle A$\odot\beta$, or of halfe the arch Aβ which is the one halfe baſe : and A*c* ſhall be the Tangent of the angle APγ, or of halfe the angle A$\odot\gamma$, or of halfe the arch Aγ, which is the other halfe baſe. Now ſeeing it hath been ſhewed that A*b* is the Tangent of one of the halfe

CHAP.*6*   *The second Booke.*   79

halfe bafes, and A*c* the Tangent of the other halfe bafe, and that A*d* is the Tangent of half the difference of the legs, and A*e* the Tangent of halfe the fumme of the legs. I fay, that

As A*b* *the Tangent of the true halfe bafe, is to* A*e* *the Tangent of the halfe fumme of the legs :*

So is A*d* *the tangent of the halfe difference of the legs, to* A*c* *the tangent of the altern halfe bafe.*

Or contrariwife, by making the true bafe of the alterne.

As A*c* *the tangent of the true halfe bafe, is to* A*e* *the Tangent of the halfe fumme of the legs :*

So is A*d* *the tangent of the halfe difference of the legs, to* A*b* *the tangent of the alterne halfe-bafe.*

Which I proue thus : If the points *bcde* be in the fame circle, then as A*b* is to A*e*, fo is A*d* to A*c*. And contrariwife, as wee faid euen now (by *36* Prop. 3 and 16 Prop. 6. *Eucl.*) But the poynts *bcde* fall in the fame circle. For the fhadow of any circle defcribed in the fuperficies of a fphære comming from a lightfome body in the fame fuperficies which is not in the circumference of the circle, maketh a circle perfectly round in the plaine perpendicular to the right line which goeth from the lightfome body by the center of the fphære, as it is manifeft out of the Optickes, & making of the *Aftrolabe* : and by *Apollonius* in his 1 book of Conick fections prop.5. But here the circle δβγε is defcribed in the fuperficies of the Sphære, and the lightfome body P is without the circumference of the circle, and the right line that goeth from the fame by the center (that is P⊙A is perpendicular to the plaine) therefore the fhadow of that circle which falleth vpon the

E 4   poynts

80   *The second Booke.*   CHAP.*6*

poynts *dbce*, is neceffarily circular, and perfectly round. Therefore

     As Ab is to Ae, fo is Ad to Ac.

And contrariwife, that is,

     *As the tangent of the true halfe base, is to the
tangent of the halfe fumme of the legs :*

     *So is the tangent of the halfe difference of the
legs, to the tangent of the alterne halfe bafe.* And
by confequent, the *Differential of the true halfe
bafe, fubtracted out of the fumme of the Differentials of the halfe fumme and halfe difference of the
legs, is equall to the Differentiall of the Alterne
halfe bafe,* which things we vndertooke to de-
monftrate.

7   *Therefore three fides of a Sphæicall triangle
being giuen, any one of the angles is had three
wayes.*

8   The firft way is, *That you make any fide the
bafe (efpecially that which commeth neereft a quadrant) then*

     *Adde halfe the bafe and halfe the difference of
the legs together, and to the Logarithme thereof
adde the Logarithme of the difference of them;
out of which fumme fubduct the fumme of the Logarithme of the legs : and the halfe of the remainder is the Logarithme of an arch, which being
doubled is the verticall angle. And fo the reft.*

     As of the triangle PZS, let the fides PZ
be giuen 34 deg. and ZS 47 deg. and SP 69
degr. let the an-
gles bee fought
out, and firft the
angle PZS cō-
ming neereft a
quadrant, which
SP 69 deg (that
is, the fide nereft

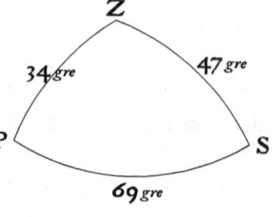

            a qua-

CHAP.*6.*   *The second Booke.*   81

a quadrant) ſubtendeth. Therefore let this
SP *69* be made the baſe : then
Adde $\frac{1}{2}$ the baſe PS —— 34 deg. 36
To $\frac{1}{2}$ the difference of $\Big\}$ *6* deg. 30
the legs PZ and ZS —
The ſumme is 41 deg. 6. the Logar.   421504
The differ. is 28 deg. 6. the Logar.   756147

         The ſumme is   **+** 1177651

Adde the Logar. of the
leg PZ 34 deg. 581260 $\Big\}$ ſumme **+** 894118
To the Log. of the legg.
ZS 47 deg. **+** 312858

    Subtract the ſame out $\Big\}$ the reſt is 283533
    of the former ſumme,
       whereof take the halfe **+** 141766

W^{ch} is the Logar. of *60* deg. 12 24 $\frac{1}{2}$
and being doubled is 120 deg. 24 49. the verticall angle PZS that was ſought for.

    No otherwiſe you may (if you will) finde
out the other angles : but they ſhall be found
more eaſily by 9 *Chap. 5.* of this booke, becauſe
(by 2 *Sent. 3 chap.*) they are of a certaine and
knowne kinde.

The ſecond way is, *that any ſide (eſpecially that* 9
which is neereſt a quadrant) being made the baſe,
you adde halfe the baſe, and halfe the ſumme of
the legs together, and to the Logarithme thereof
adde the Logarithme of the difference of them :
out of which ſumme ſubduct the ſumme of the Lo-
garithme of the legges, and the baſes of the
remainder is the Antilogarithme of an arch, which
being doubled, is the verticall angle, and ſo the
reſt.
As of the ſame Triangle PZS
Adde halfe the baſe PS. 34 deg. 36.

      E 5   To

82   *The second Booke.*   CHAP.*6*

To halfe the sum of the ⎱ 40 deg. 36.
legs PZ, and ZS.────── ⎰

The summe is 75 deg 6. the Logar.   2258295
The Differ. is 6 deg. 6 the Logar.   34668

The summe is ⎯⎯ $+$ 2292963

Adde the Logar. of the ⎫
leg. PZ 34 deg. 58 12 61 ⎬
To the Log. of the leg ⎬ summe is $+$ 894119
ZS. 47 deg. $+$ 312858 ⎭

Subtract the same out ⎫
of the former summe, ⎭ The rest is 1398844

whereof take the halfe $—+$ 699422

W^{ch} is the Antilog. of 60 deg 12. 24 $\frac{1}{2}$
and being doubled is 120 deg. 24. 49 the verticall angle PZS sought for.

The other angle, although you may finde
after this manner, yet you shall finde them
more easily by 9 *Chap* 5 of this booke. For by
the second sentence of the third Chap. they
are of a knowne kinde.

10   *The third way is, that any side being put for the
base, you adde the Differential of halfe the summe
of the legs, to the Differentiall of halfe the difference of the legs, and subtract from the product
the Differentiall of the true halfe base, and there
shall come thereof the Differential of the alterne
halfe base. The summe of which halfe bases is the
greater case, and the difference the lesse case, di-
stinguishing two right-angled triangles, which do
make knowne both their owne parts, and all the
parts of the triangle propofed (by 9 chap. 4 and 8
chap. 5. of this booke.)*

As the sides of the triangle propounded
PZS being giuen, as before, let the angles at
the base ZPS, and ZSP be sought for.

halfe

CHAP.*6*.   *The second Booke*.   83

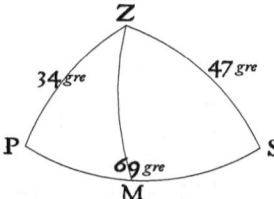

Half the sum⌉
of the leggs ⎬40 deg 30. the Differ. 157730
PZ, & ZS is ⌋

Half the dif-⌉
ference of   ⎬ 6 deg. 30. the Differ 2172121
them is ——⌋

W ᵈ added together, bring forth   2329851

From which subtract⌉
the differ : of halfe ⎬ viz. 34. deg. 30. 375012
the true base PZ. ⌋

     And there will remaine   ✛ 1954839
Which is the Differential of 8 deg. 3. 31. halfe
the alterne base.

     Adde therefore together the two halfe ba-
fes, to wit,

     Halfe the true base 34 deg. 30. 0.
and half the altern base 8 deg. 3. 31.

     And they make —— 42 deg. 33. 31. the
greater case MS.

Take the one out ⌉
     of the other, ⎬ refts 26 deg. 26. 29. the
lefse case PM.   ⌋

     Therefore by helpe of thefe cafes, you
haue now two triangles right angled at M,
that is PMZ, and SMZ, which do lay open
both the perpendicular ZM, and the verticall angles PZM and SZM : or (if you will)
PZS (by 9 *Chap*. 4 and 8 *Chap*. 5 of this book)
But thefe things being omitted, let vs return
                                                         to

84   *The second Booke.*   CHAP.*6*

to the angles of the bafe ZPS, ZSP which
were fought for.

Adde the Differential of the ⎫
cafe PM. 26, 26 29 already ⎬ —+698552
found (by 9 Sect. 4 Chap) ⎭

To the Differentiall of the ⎫
complement of PZ, which is ⎬ —— 393771
56 degrees ——— ⎭   ————

There will come forth —+304781

Which is the Logarithme of the complement of the angle ZPS, which complement
is 47 deg. 36. 1.

Likewife,

Adde the Differentiall of the ⎫
greater cafe SM 42 de. 33. 31 ⎬ —+85324
already found by the 9 Sect. ⎭

To the Differentiall of the ⎫
complement of SZ, which is ⎬ —+69870
43 degrees ——— ⎭   ————

There will come forth —+ 155194

Which is the Logarithme of the complement of the angle ZSP, which complement
is 58 deg. 53. 55.

But here remember, that not the parts PZ
34, and ZPS, or SZ 47, and ZSP, but their
complements, that is, 56 degr. and 47, 36, 1.
and 43 deg. and 58 53, 55. are here called circular parts, (by 2 Chap. 4. of this booke)
Therefore the true angle fought for ZPS, is
42, 29, 59.

And ZSP is 31, 6, 5.

As it alfo manifeft by 8 Sect. Chap 5 of this
booke.

Another example of the fame triangle.

T he fame triangle PZS being placed o-
therwife,

CHAP.**6.**   *The second Booke.*   **85**

therwife, let ZS be the bafe, and the fides being giuen, as before, let the angle PZS bee
fought for. Therefore

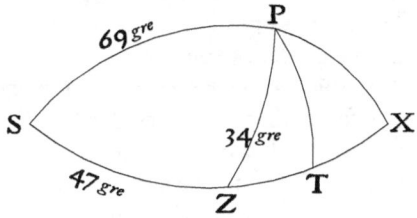

Halfe the fum
of the legs S P
and PZ is —— } 51 deg.36 the Differ—228865

Half the diffe-
rence of them
is —— } 17 de.36.the Diff.+ 1154234

Which added together,
bring forth the Differ. } —— + 925369

From wch fubtract the
Differentiall of halfe
the true bafe SZ, viz. of } 23 de.30 + 832840

And there will remaine —— + 92529
W^{ch} is the Differential of 42 deg. 2́1. 1́1.halfe
the alterne bafe.

Adde therefore together the two halfe bafes, to wit,

Halfe the true bafe 23 deg. 36, 6́.
And half the alterne bafe 42 deg. 2́1, 1́1,

And there wil come forth 65 deg. 5́1, 1́1, the
greater cafe ST.
Take the difference of them 18 deg. 5́1, 1́1, the
leffe cafe TX. or TZ.
Adde therefore the Diffe-
rentiall of the fame TZ,
18, 5́1, 1́1, viz, —— } —— + 1074520

To

86   *The ſecond Booke.*   Cʜᴀᴘ.*6*

To the differentiall of the complement of ZP, which is 56 degrees $\rbrace$ ———— 393771

And from thence will ariſe —— $+$ 680749

Which is the Logarithme of the complement of the angle PZT. 59 deg 35. 11

Of which angle PZT, ſeeing the angle ſought for PZS is the remainder to a ſemi-circle (which alwayes happeneth when the alterne baſe is greater then the true) the angle PZS muſt needs be 120 degrees, 24. 49, otherwiſe if the true baſe exceed the alterne baſe, the angles PZT, and PZS ſhall bee all one.

An Admonition.

YOu haue now three true wayes to finde out the angles by the ſides giuen by euery one whereof three ſeuerall queſtions of this and any other triangle are reſolued.

For by *the eleuation of the pole, the height of the Sunne, and declination of the ſunne* being giuen, they that doubt are ſatiſfied concerning the queſtion : whereby either

1 The Azimuth of the Sunne,

2 The angle of the ſituation and poſition of the Sunne : or

3 The houre of the day is demanded.

Hitherto wee haue found the angles by the ſides.

It remaineth to finde the ſides by the angles.

11   *In any Sphæricall Triangle the ſides may bee changed into angles, and the angles into ſides : yet taking firſt for any one angle and his ſubtending ſide, the remainders of them to a ſemicircle.*

For

CHAP.*6.*   *The second Booke.*   87

For example sake.

Let QRT be a tri-
angle, whose angles
let bee Q 47, R III,
and T 34.

Let vs first take for
any angle, as for R III,
his remainder to a se-
micircle, which is 69
degrees.

I say that these an-
gles 47ᵈ. 69ᵈ. 34ᵈ.
may bee chan-
ged into sides, &
the triangle
PZS going a-
fore, and heere
now againe ex-
pressed shall be
made.

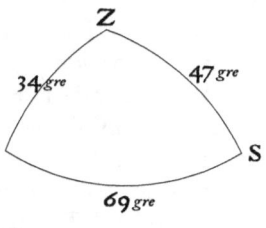

Wherein PZ is 34 degrees,

ZS is 47 degrees,

And PS is 69 degrees,

As also of the angles of this repeated triangle shall mutually bee made the sides of
the other:

For the angle of this triangle ZSP 31, 6. 5̈.
is the side of the other triangle QR.

And the angle of this ZPS 42, 29́, 59̈. is
the side of the other RT.

And of the third angle of this triangle
which is SZP 120, 24́, 49̈. the remainder to
a semicircle that is 59, 35́, 11̈, is the side QT
of the other triangle.

The demonstration whereof *Bartholomew
Pitiscus,* *Adrian Metius* and others set downe,
therefore I thinke it no need to repeat the
                                        same

fame in this fhort Treatife.

12   *From whence it followeth, that the three angles of a Sphæricall triangle being giuen, the fides are found by an eafie conuerfion.*

As in the former triangle QRT, let the angles be giuen Q 47, R 111, and T 34, let the fides be fought.

For any one angle, for example fake,(as be-fore) for R 111, let the remainder thereof to a femicircle 69 degrees be taken.

Then 47, 69, 34 being fet for fides (as was done in the former triangle PZS, by any of the three wayes aboue written, feeke his angles, and you fhall finde,

Againft the fide 47, the angle 42. 29̃. 59̃.
And againft the fide 34 the angle 31, 6̃. 5̃.
And againft the fide 69 (which wee put for 111) you fhall finde the angle 120, 24̃, 49̃. Therfore in the triangle propounded, QRT.

For the fide RT fubtending the angle Q 47, fet downe 42 29̃ 59̃.

And for the fide QR fubtending the an-gle T 34, fet downe 31. 6̃. 5̃.

But for the fide QT fubtending the angle R 111. fet downe 59, 35̃, 11̃.

Which are the remainder of 120 deg. 24̃. 49̃. to a femicircle, becaufe before you tooke for 111 his remainder to a femicircle, that is 69. And fo by conuerfion you fhal finde the fides by the angles.

An Admonition.

OVt of this finding of the fides by the an-gles giuen, three diuers queftions of this and any other triangle whatfoeuer are refol-ued.

As in the triangle PZS out of *the houre of the day, the azimuth of the Sunne, and the angle*

of

CHAP.*6.*   *The second Booke.*   89

of the site or position of the Sunne, this propofition going before, giueth fatisfaction to the
queftion, whereby either

   1 The height of the pole,
   2 The height of the Sunne, or
   3 The declination of the Snnne is demanded.

Therefore (out of the 8 fect. of the former
5 chap. and the 7 & 13 fect.of this booke) you
haue the folution of 60 feuerall queftions,
wch fall into any triangle : neither can there
any moe varieties then thefe arife out of the
manifold compofition of any three parts.

You haue therfore a perfect & abfolute doctrine of triangles, as well *Sphæricall* as *Plaine.*

THE CONCLVSION.

NOw *therfore it hath been fufficiently fhewed
that there are Logarithmes, what they are,
and of what vfe they are : for with helpe of them
we haue both demonftratiuely fhewed and taught
by examples of both kindes of Trigonometrie, that
the Arithmeticall folution of any Geometricall
queftion may moft readily bee performed without
trouble of Multiplication, Diuifion, or extraction
of roots. You haue therfore the admirable Table of
Logarithmes that was promifed, together with the
moft plentifull vfe thereof, which if (to you
of the learneder fort) I fhall by your letters vnderftand to be acceptable to you, I fhall be incouraged
to fet forth alfo the way to make the Table. In the
meane time, make vfe of this fhort Treatife, and
giue all praife and glory to God the high Inuenter
and guider of all good workes.*

The end of the Treatife.
*Now followeth the Table or Canon of
Logarithmes.*

Deg. 0   +|—

mi	Sines.	Logarith.	Differen.	Logarith.	Sines.	
0	0	*Infinite.*	*Infinite.*	.0	1000000.0	60
1	291	3142567	8142568	.1	1000000.0	59
2	582	7449419	7449421	.2	999999.8	58
3	873	7043952	7043956	.4	999999.6	57
4	1164	6756275	6756274	.7	999999.3	56
5	1454	6533131	6533130	1.1	999998.9	55
6	1745	6350810	6350808	1.6	999998.6	54
7	2036	6196659	6196657	2.2	999998.0	53
8	2327	6063128	6063126	2.8	999997.4	52
9	2618	5945345	5945342	3.5	999996.7	51
10	2909	5839986	5839814	4.3	999995.9	50
11	3280	5744676	5744671	5.2	999995.0	49
12	3491	5657665	5657658	6.2	999994.0	48
13	3781	5577622	5577615	7.3	999992.8	47
14	4072	5503514	5503506	8.4	999991.7	46
15	4363	5434522	5434513	9.6	999990.5	45
16	4654	5369984	5369973	10.9	999989.2	44
17	4945	5309360	5309348	12.3	999987.8	43
18	5236	5252202	5252188	13.8	999986.3	42
19	5527	5198136	5198120	15.4	999984.7	41
20	5818	5146843	5146336	17.0	999983.1	40
21	6109	5098054	5098045	18.7	999981.3	39
22	6399	5051534	5051514	20.5	999979.5	38
23	6690	5007083	5007060	22.4	999977.6	37
24	6981	4964524	4964499	24.4	999975.6	36
25	7272	4923703	4923676	26.5	999973.6	35
26	7563	4884483	4884454	28.7	999971.4	34
27	7854	4846743	4846712	30.9	999969.1	33
28	8145	4810376	4810343	33.2	999966.8	32
29	8436	4775286	4775250	35.4	999964.4	31
30	8726	4741385	4741347	38.1	999961.9	30

Min.

Deg. 89

Deg. 0 $+|-$

mi	Sines.	Logarith.	Differen.	Logarith.	Sines.	
30	8726	4741385	4741347	38.1	999961.9	30
31	9017	4708596	4708555	40.7	999959.3	29
32	9308	4676848	4676805	43.4	999956.6	28
33	9599	4646077	4646031	46.1	999953.9	27
34	9890	4616225	4616176	48.9	999951.1	26
35	10181	4587239	4587187	51.8	999948.2	25
36	10472	4559069	4359014	54.8	999945.2	24
37	10763	4531671	4531613	57.9	999942.1	23
38	11054	4505004	4504943	61.1	999938.9	22
39	11344	4479030	4478965	64.4	999935.7	21
40	11635	4453713	4453645	67.7	999932.3	20
41	11926	4429022	4428950	71.1	999928.9	19
42	12217	4404925	4404850	74.6	999925.4	18
43	12508	4381396	4381318	78.2	999921.8	17
44	12799	4358408	4358326	81.9	999918.1	16
45	13090	4335936	4335850	85.7	999914.3	15
46	13380	4313958	4313868	89.6	999910.5	14
47	13671	4292453	4292360	93.5	999906.5	13
48	13962	4271401	4271304	97.5	999902.5	12
49	14253	4250783	4250682	101.6	999898.4	11
50	14544	4230583	4230477	105.8	999894.2	10
51	14835	4210781	4210671	110.1	999890.0	9
52	15126	4191364	4191250	114.5	999885.6	8
53	15416	4172317	4172198	118.9	999881.1	7
54	15707	4153627	4153504	123.4	999876.6	6
55	15998	4135279	4135151	128.0	999872.0	5
56	16289	4117263	4117130	132.7	999867.3	4
57	16580	4100664	4100527	137.5	999862.5	3
58	16871	4082175	4082032	142.4	999857.7	2
59	17162	4065082	4064935	147.3	999852.7	1
60	17452	4048276	4048124	152.3	999847.7	0

Min.

Deg. 89

Deg. 1   +|−

m.	Sines.	Logarith.	Differen.	Logarit.	Sines.	
0	17452	4048276	4048124	152	999848	60
1	17743	4031748	4031591	157	999843	59
2	18034	4015490	4015327	163	999837	58
3	18325	3999492	3999324	168	999832	57
4	18615	3983745	3983571	173	999827	56
5	18907	3968242	3968063	179	999811	55
6	19197	3952976	3952792	184	999816	54
7	19488	3937941	3937751	190	999810	53
8	19779	3923127	3922932	196	999804	52
9	20070	3908531	3908329	201	999799	51
10	20361	2894144	3893937	207	999793	50
11	20652	3879961	3879748	213	999787	49
12	20942	3865977	3865757	219	999781	48
13	21233	3852186	3851960	225	999774	47
14	21524	3838582	3838351	232	999768	46
15	21815	3825161	3824923	238	999762	45
16	22106	3811918	3811674	244	999756	44
17	22396	3798848	3798597	251	999749	43
18	22687	3785947	3955690	257	999742	42
19	22978	3773210	3772946	264	999736	41
20	23269	3760634	3760363	271	999730	40
21	23568	3748213	3747936	278	999722	39
22	23850	3735946	3735661	284	999715	38
23	24141	3723827	3723535	291	999708	37
24	24432	3711853	3711555	299	999701	36
25	24723	3700021	3699715	306	999694	35
26	25014	3688327	3688014	313	999687	34
27	25304	3676769	3676449	320	999680	33
28	25595	3665343	3665015	328	999672	32
29	25886	3654045	3653710	335	999665	31
30	26177	3642875	3642532	343	999657	30

Min.

Deg. 88

Deg. 1 +|−

m.	Sines.	Logarith.	Differen.	Logarit.	Sines.	
30	26177	3642875	3642532	343	999657	30
31	26468	3631827	3631477	350	999650	29
32	26758	3620901	3620543	368	999642	28
33	27049	3610092	3609726	366	999634	27
34	27340	3599400	3599026	374	999626	26
35	27631	3588821	3588439	382	999618	25
36	27922	3578352	3577962	390	999610	24
37	28212	3567992	3567593	398	999602	23
38	28503	3557730	3557331	406	999594	22
39	28794	3547589	3547174	415	999585	21
40	29085	3537541	3537118	423	999577	20
41	29375	3527593	3527162	431	999568	19
42	29666	3517744	3517304	440	999560	18
43	29957	3507991	3507542	449	999551	17
44	30248	3498332	3497874	458	999542	16
45	30538	3488765	3488299	466	999534	15
46	30829	3479289	3478814	475	999525	14
47	31120	3469903	3469418	484	999516	13
48	31411	3460604	3460110	493	999507	12
49	31701	3451390	3450887	503	999497	11
50	31992	3442261	3441749	512	999488	10
51	32283	3433214	3432693	521	999479	9
52	32574	3424248	3423718	530	999469	8
53	32864	3415363	3414823	540	999460	7
54	33155	3407655	3407105	550	999450	6
55	33446	3397825	3397265	560	999440	5
56	33737	3389170	3388601	569	999431	4
57	34027	3380589	3380010	579	999421	3
58	34318	3372082	3371493	589	999411	2
59	34609	3363646	3363047	599	999401	1
60	34899	3354528	3354672	609	999391	0

Min.

Deg. 88

Deg. 2   +|—

m.	Sines.	Logarith.	Differen.	Logarit.	Sines.	
0	34899	3355282	3354672	609	999391	60
1	35190	3346986	3346366	620	999381	59
2	35481	3338759	3338129	630	999370	58
3	35772	3330599	3329959	640	999360	57
4	36062	3322506	3321855	650	999349	56
5	36353	3314477	3313816	661	999339	55
6	36644	3306513	3305841	672	999328	54
7	36934	3298611	3297928	682	999318	53
8	37225	3290771	3290078	693	999307	52
9	37516	3282992	3282288	704	999296	51
10	37806	3275274	3274559	715	999282	50
11	38097	3267615	3266889	726	999270	49
12	38388	3260014	3259287	737	999263	48
13	38678	3252471	3251722	748	999252	47
14	38969	3244984	3244224	760	999240	46
15	39260	3237553	3236781	771	999229	45
16	39550	3230176	3229393	783	999217	44
17	39841	3222854	3222060	794	999206	43
18	40132	3215585	3214779	806	999195	42
19	40422	3208369	3207552	818	999183	41
20	40713	3201204	3200375	830	999171	40
21	41004	3194091	3193249	841	999159	39
22	41294	3187028	3186174	853	999147	38
23	41585	3180014	3179149	865	999135	37
24	41876	3173049	3172172	878	999123	36
25	42166	3166133	3165243	890	999110	35
26	42457	3159264	3158362	902	999098	34
27	42748	3152442	3151528	914	999086	33
28	43038	3145667	3144740	927	999073	32
29	43329	3138937	3137997	940	999061	31
30	43619	3132252	3131300	952	999048	30

Min.

Deg. 87

Deg. 2   +|−

m.	Sines.	Logarith.	Differen.	Logarit.	Sines.	
30	43619	3132252	3131300	952	999048	30
31	43910	3125612	3124647	965	999035	29
32	44201	3119016	3118038	978	999023	28
33	44491	3112463	3111472	991	999010	27
34	44782	3105952	3104948	1004	998997	26
35	45072	3099484	3098467	1017	998984	25
36	45363	3093058	3092028	1030	998971	24
37	45654	3086672	3085629	1043	998957	23
38	45944	3080328	3079271	1056	998944	22
39	46235	3074023	3072953	1070	998931	21
40	46525	3067758	3066674	1083	998917	20
41	46816	3061532	3060435	1097	998904	19
42	47106	3055344	3054233	1111	998890	18
43	47397	3049195	3048070	1124	998876	17
44	47688	3043083	3041945	1138	998862	16
45	47978	3037009	3035857	1152	998848	15
46	48269	3030971	3029805	1166	998834	14
47	48559	3024970	3023790	1180	998820	13
48	48850	3019005	3017810	1194	998806	12
49	49140	3013075	3011866	1209	998792	11
50	49431	3007180	3005950	1223	998777	10
51	49721	3001319	3000082	1238	998763	9
52	50012	2995493	2994241	1252	998749	8
53	50302	2989701	2988435	1267	998734	7
54	50593	2983942	2982661	1281	998719	6
55	50883	2978216	2976920	1296	998704	5
56	51174	2972524	2971212	1311	998690	4
57	51464	2966863	2965537	1326	998675	3
58	51755	2961233	2959892	1341	998660	2
59	52045	2955636	2954280	1356	998645	1
60	52336	2950071	2948699	1372	998629	0

Min.

Deg. 87

Deg. 3   +|−

m.	Sines.	Logarith.	Differen.	Logarit.	Sines.	
0	52336	2950071	2948699	1371	998629	60
1	52626	2944535	2943149	1387	998614	59
2	52927	2939030	2937629	1402	998599	58
3	53207	2933556	2932139	1418	998583	57
4	53498	2928112	2926671	1433	998568	56
5	53788	2922697	2921249	1449	998552	55
6	54079	2917311	2915847	1464	998537	54
7	54369	2911954	2910475	1480	998521	53
8	54660	290662	2905131	1496	998505	52
9	54950	2901327	2899815	1512	998489	51
10	55241	2896056	2894528	1518	998473	50
11	55531	2890812	2889267	1544	998457	49
12	55822	2885595	2884035	1560	998441	48
13	56112	2880406	2878829	1577	998424	47
14	56402	2875243	2873650	1593	998408	46
15	56693	2870107	2868497	1610	998392	45
16	56983	2864997	2863371	1626	998375	44
17	57274	2859914	2858271	1643	998359	43
18	57564	2854857	2853198	1659	998342	42
19	57854	2849825	2848148	1676	998325	41
20	58145	2844818	2843125	1693	998308	40
21	58435	2839835	2838125	1710	998291	39
22	58726	2834878	2833151	1727	998274	38
23	59016	2829946	2828201	1744	998257	37
24	59306	2825038	2823276	1762	998240	36
25	59597	2820153	2828375	1779	998222	35
26	59887	2815293	2813497	1796	998205	34
27	60177	2810456	2808642	1814	698188	33
28	60468	2805643	2803811	1831	998170	32
29	60758	2800852	2799003	1849	998152	31
30	61048	2796085	2794218	1867	998135	30

Min.

Deg. 86

Deg. 3   +|−

m.	Sines.	Logarith.	Differen.	Logarit.	Sines.	
30	61048	2796085	2794218	1867	998135	30
31	61339	2791340	2789455	1885	998117	29
32	61629	2786618	2784715	1903	998099	28
33	61920	2781918	2779998	1921	998081	27
34	62210	2777241	2775302	1939	998063	26
35	62500	2772585	2770628	1957	998045	25
36	62790	2767950	2765975	1975	998027	24
37	63081	2763337	2761344	1993	998008	23
38	63371	2758746	2756734	2012	997990	22
39	63661	2754175	2752145	2030	997972	21
40	63952	2749626	2747577	2049	997953	20
41	64242	2745097	2743029	2068	997934	19
42	64532	2740588	2738502	2086	997916	18
43	64823	2736100	2733995	2105	997897	17
44	65113	2731632	2729508	2124	997876	16
45	65403	2727184	2725041	2143	997859	15
46	65693	2722756	2720594	2162	997840	14
47	65984	2718348	2716166	2181	997821	13
48	66274	2713958	2711757	2201	997801	12
49	66564	2709588	2707368	2220	997782	11
50	66854	2705237	2702998	2240	997763	10
51	67145	2700906	2698646	2259	997743	9
52	67435	2696592	2694314	2279	997724	8
53	67725	2692298	2689999	2298	997704	7
54	68015	2688022	2685703	2318	997684	6
55	68305	2683764	2681425	2338	997664	5
56	68596	2679524	2677166	2358	997645	4
57	68886	2675303	2672924	2378	997625	3
58	69176	2671098	2668700	2398	997604	2
59	69466	2666913	2664494	2419	997584	1
60	69756	2662744	2660305	2439	997564	0

Min.

F   Deg. 86

Deg. 4   +|−

m.	Sines.	Logarith.	Differen.	Logarit.	Sines.	
0	69756	2662744	2660305	2439	998564	60
1	70047	2658593	2656133	2459	997544	59
2	70337	2654459	2651979	2480	997523	58
3	70627	2650342	2647841	2500	997503	57
4	70917	2646242	2643721	2521	997482	56
5	71207	2642159	2639617	2542	997461	55
6	71497	2638093	2635530	2563	997441	54
7	71788	2634043	2631459	2583	997420	53
8	72078	2630009	2627405	2604	997399	52
9	72368	2625992	2623367	2625	997378	51
10	72658	2621991	2619345	2646	997357	50
11	72948	2618007	2615339	2668	997336	49
12	73238	2614038	2611349	2689	997314	48
13	73528	2610084	2607373	2710	997293	47
14	73818	2606146	2603415	2732	997272	46
15	74108	2602224	2599471	2753	997250	45
16	74398	2598318	2595542	2775	997229	44
17	74689	2594426	2591629	2797	997207	43
18	74979	2590550	2587731	2819	997185	42
19	75269	2586688	2583848	2841	997163	41
20	75559	2582842	2579980	2863	997141	40
21	75849	2579011	2575126	2885	997119	39
22	76139	2575194	2572287	2907	997097	38
23	76429	2571392	2568473	2929	997075	37
24	76718	2567604	2564653	2952	997053	36
25	77009	2563831	2560857	2974	997030	35
26	77299	2560072	2557076	2996	997008	34
27	77589	2556327	2553308	3019	996985	33
28	77879	2552597	2549555	3042	996963	32
29	78169	2548880	2545815	3065	996940	31
30	78459	2545177	2542089	3087	996917	30

Min.

Deg. 85

Deg. 4   + | −

m.	Sines.	Logarith.	Differen.	Logarit.	Sines.	
30	78459	2545177	2542089	3087	996917	30
31	78749	2541488	2538377	3110	996894	29
32	79039	2537812	2534678	3133	996871	28
33	79329	2534150	2530993	3156	996848	27
34	79619	2530501	2527322	3180	996825	26
35	79909	2526866	2523663	3203	996802	25
36	80199	2523244	2520018	3226	996779	24
37	80489	2519635	2516386	3250	996755	23
38	80779	2516040	2512767	3273	996732	22
39	81069	2512457	2509160	3297	996708	21
40	81359	2508887	2505566	3321	996685	20
41	81649	2505330	2501985	3344	999661	19
42	81938	2501785	2498417	3368	996637	18
43	82228	2498253	2494861	3392	996613	17
44	82518	2494734	2491318	3416	996589	16
45	82808	2491227	2487787	3440	996565	15
46	83098	3487733	2484268	3465	996541	14
47	83388	2484250	2480761	3489	996517	13
48	83678	2480780	2477267	3513	996493	12
49	83968	2477322	2473784	3537	996468	11
50	84257	2473806	2470314	3562	996444	10
51	84547	2470442	2466855	3587	996419	9
52	84837	2467020	2463408	3612	996395	8
53	85127	2463609	2459973	3636	996370	7
54	85417	2460210	2456549	3661	996345	6
55	85707	2456823	2453136	3686	996320	5
56	85996	2453447	2449736	3711	996295	4
57	86286	2450083	2446346	3737	996270	3
58	86576	2446730	2442968	3762	996245	2
59	86866	2443388	2439601	3787	996220	1
60	87156	2440058	2436245	3813	991695	0

Min.

F 2   **Deg. 85**

Deg. 5   +|−

m.	Sines.	Logarith.	Differen.	Logarit.	Sines.	
0	87156	2440058	2436245	3813	996195	60
1	87445	2436738	2432900	3838	996169	59
2	87735	2433430	2429566	3864	996144	58
3	88025	3430133	2426243	3889	996118	57
4	88315	2426847	2422932	3915	996093	56
5	88604	2423571	2419630	3941	996067	55
6	88894	2420306	2416340	3967	996041	54
7	89184	2417052	2413059	3993	996015	53
8	89474	2413809	2409790	4019	995989	52
9	89763	2410576	2406531	4045	995963	51
10	90053	2407354	2403282	4071	995937	50
11	90343	2404142	2400045	4098	995910	49
12	90632	2400941	2396817	4124	995884	48
13	90922	2397749	2393599	4150	995858	47
14	91212	2394568	2390391	4177	995831	46
15	91502	2391398	2387194	4204	995805	45
16	91791	2388237	2384007	4230	995778	44
17	92081	2385087	2380829	4257	995751	43
18	92371	2381946	2377661	4284	995724	42
19	92660	2378815	2374504	4311	995698	41
20	92950	2375694	2371356	4339	995670	40
21	93239	2372583	2368217	4365	995644	39
22	93529	2369482	2365089	4393	995616	38
23	93819	2366390	2361969	4420	995589	37
24	94108	2363308	2358860	4448	995562	36
25	94398	2360235	2335760	4475	995535	35
26	94687	2357172	2352669	4503	995507	34
27	94977	2354119	2349588	4531	995479	33
28	95267	2351075	2346516	4558	995452	32
29	95556	2348040	2343453	4586	994424	31
30	95846	2345014	2340400	4614	995396	30

Min.

Deg. 84

Deg. 5   +|−

m.	Sines.	Logarith.	Differen.	Logarit.	Sines.	
30	95846	2345014	2340400	4614	995396	30
31	96135	2341998	2337356	4642	995368	29
32	96425	2338991	2334320	4670	995340	28
33	96714	2335993	2331294	4699	995312	27
34	97004	2333004	2328277	4727	995284	26
35	97293	2330023	2325268	4755	995256	25
36	97583	2327052	2322269	4784	995227	24
37	97872	2324090	2319278	4812	995199	23
38	98162	2321137	2316296	4841	995170	22
39	98451	2318192	2313322	4870	995142	21
40	98741	2315256	2310357	4899	995113	20
41	99030	2312220	2307401	4928	995084	19
42	99320	2309410	2304453	4957	995056	18
43	99609	2306500	2301514	4986	995027	17
44	99899	2303598	2298584	5015	994998	16
45	100188	2300706	2295661	5041	994968	15
46	100477	2297821	2292748	5073	994939	14
47	100767	2294945	2289842	5103	994910	13
48	101056	2292077	2286945	5132	964881	12
49	101346	2289217	2284055	5162	994851	11
50	101635	2286366	2281172	5192	994822	10
51	101924	2283523	2278301	5221	994792	9
52	102214	2280688	2275437	5251	994762	8
53	102503	2277861	2272580	5281	994733	7
54	102793	2275042	2269231	5311	994703	6
55	103082	2272231	2266890	5341	994673	5
56	103371	3269428	2264057	5371	994643	4
57	103661	2266633	2261232	5402	994613	3
58	103950	2263846	2258414	5432	994583	2
59	104239	2261066	2255604	5463	994552	1
60	104528	2258295	2252802	5493	994522	0

Min.

F 3   Deg. 84

Deg. 6   +|−

m.	Sines.	Logarith:	Differen.	Logarit.	Sines.	
0	104528	2258295	2252802	5493	994522	60
1	104818	2255531	2250007	5524	994491	59
2	105107	2252775	2247121	5554	994461	58
3	105396	2250027	2244441	5585	994430	57
4	105686	2247286	2241670	5616	994400	56
5	105975	2244553	2238905	5647	994369	55
6	106264	2241827	2236149	5678	994338	54
7	106553	2239109	2233400	5709	994307	53
8	106843	2236398	2230658	5740	994276	52
9	107132	2233695	2227923	5772	994245	51
10	107421	2230999	2225196	5803	994214	50
11	107710	2228310	2222476	5835	994182	49
12	107999	2225629	2219763	5866	994151	48
13	108289	2222954	2217057	5898	994119	47
14	108578	2220288	2214358	5930	994088	46
15	108867	2217628	2211667	5961	994056	45
16	109156	2214976	2208983	5993	994925	44
17	109445	2212331	2206305	6025	993993	43
18	109734	2209692	2203635	6057	993961	42
19	110023	2207061	2200972	6089	993929	41
20	110313	2204437	2198315	6122	993897	40
21	110602	2201819	2195665	6154	993865	39
22	110891	2199209	2193023	6186	993833	38
23	111180	2196605	2190386	6219	993800	37
24	111469	2194009	2187757	6251	993768	36
25	111758	2191419	2185134	6284	993735	35
26	112047	2188835	2182518	6317	993703	34
27	112336	2186259	2179909	6350	993670	33
28	112625	2183689	2177306	6383	993638	32
29	112914	2181126	2174710	6416	993605	31
30	113203	2178570	2172121	6449	993572	30

Min.

Deg. 83

Deg. 6   + | −

m.	Sines.	Logarith.	Differen.	Logarit.	Sines.	
30	113203	2178570	2172121	6449	993572	30
31	113492	2176020	2169538	6482	993539	29
32	113781	2173477	2166961	6515	993506	28
33	114070	2170940	2164392	6549	993473	27
34	114359	2168410	2161828	6582	993439	26
35	114648	2165886	2159270	6616	993406	25
36	114937	2163369	2156720	6649	993373	24
37	115226	2160859	2154176	6683	993339	23
38	115515	2158354	2151637	6717	993306	22
39	115804	2155856	2149105	6751	993272	21
40	116093	2153364	2146579	6785	993238	20
41	116382	2150878	2144059	6819	993205	19
42	116671	2148399	2141546	6853	993171	18
43	116960	2143925	2139038	6887	993137	17
44	117248	2143458	2136537	6921	993103	16
45	117537	2140998	2134042	6956	993068	15
46	117826	2138543	2131553	6990	993034	14
47	118115	2136095	2129070	7025	993000	13
48	118404	2133652	2126593	7059	992966	12
49	118693	2131216	2124122	7094	992931	11
50	118982	2128785	2121657	7129	992896	10
51	119270	2126361	2119197	7164	992862	9
52	119559	2123942	2116744	9199	992827	8
53	119848	2121530	2114296	7234	992792	7
54	120137	2119123	2111854	7269	992757	6
55	120425	2116722	2109418	7304	992722	5
56	120714	2114327	2106988	7340	992687	4
57	121003	2111938	2104563	7375	992052	3
58	121292	2109555	2102144	7410	992617	2
59	121580	2107177	2099731	7446	992582	1
60	121869	2104805	2097323	7482	992546	0

Min.

F 4   **Deg. 8**

Deg. 7   +|−

mi	Sines.	Logarith:	Differen.	Logarit.	Sines.	
0	121869	2104805	2097232	7482	992546	60
1	122158	2102438	2094921	7518	992511	59
2	122447	2100078	2092524	7553	992475	58
3	122735	2097723	2090134	7589	992439	57
4	123024	2095374	2087748	7625	992404	56
5	123313	2093030	2085369	7661	992368	55
6	123601	2090692	2082995	7698	992332	54
7	123890	2088359	2080626	7734	992296	53
8	124179	2086032	2078261	7770	992260	52
9	124467	2083711	2075904	7807	992224	51
10	124756	2081394	2073551	7843	992187	50
11	125011	2079084	2071204	7880	992151	49
12	125333	2076778	2068862	7917	992115	48
13	125622	2074478	2066525	7953	992078	47
14	125910	2072184	2064193	7990	992042	46
15	125199	2069895	2061867	8027	992005	45
16	126488	2067611	2059546	8064	991968	44
17	126776	2065352	2057231	8101	991931	43
18	127065	2063059	2054920	8138	991894	42
19	127353	2060791	2052615	8176	991857	41
20	127642	2058528	2050315	8213	991820	40
21	127930	2056270	2048019	8251	991783	39
22	128219	2054018	2045729	8288	991746	38
23	128507	2051770	2043444	8326	991709	37
24	128796	2049528	2041164	8364	991671	36
25	129084	2047291	2038889	8401	991634	35
26	129373	2045059	2036619	8439	991596	34
27	129661	2042832	2034354	8477	991558	33
28	129949	2040610	2032094	8515	991521	32
29	130238	2038392	2029839	8554	991483	31
30	130526	2036180	2027589	8592	991445	30

Min.

Deg. 82

Deg. 7   +|−

mi	Sines.	Logarith.	Differen.	Logarit.	Sines.	
30	130526	2036180	2027589	8592	991445	30
31	130815	2033974	2025343	8630	991407	29
32	131103	2031772	2023103	8669	991369	28
33	131391	2029575	2020867	8707	991331	27
34	131680	2027382	2018636	8746	991292	26
35	131968	2025195	2016410	8784	991254	25
36	132256	2023012	2014189	8823	991216	24
37	132545	2020834	2011972	8862	991177	23
38	132833	2018661	2009760	8901	991138	22
39	133121	2016493	2007553	8940	991100	21
40	133410	2014330	2005351	8979	991061	20
41	133698	2012172	2003153	9018	991022	19
42	133986	2010018	2000960	9058	990983	18
43	134274	2007869	1998772	9097	990944	17
44	134563	2005724	1996588	9136	990905	16
45	134851	2003585	1994409	9176	990866	15
46	135139	2001449	1992234	9216	990827	14
47	135427	1999319	1990063	9255	990787	13
48	135716	1997193	1987898	9295	990748	12
49	136004	1995072	1985737	9335	990708	11
50	136292	1992955	1983580	9375	990669	10
51	136580	1990843	1981428	9415	990629	9
52	136868	1988736	1679280	9455	990589	8
53	137156	1986633	1977137	9495	990549	7
54	137445	1984534	1974998	9536	990509	6
55	137733	1982440	1972864	9576	990169	5
56	138021	1980350	1970734	9617	990429	4
57	138309	1978265	1968603	9657	990389	3
58	138597	1976184	1966486	9698	990349	2
59	128885	1974108	1964369	9739	990308	1
60	139173	1972036	1962257	9780	990268	0

Min.

F 5   **Deg. 82**

Deg. 8   +|−

m.	Sines.	Logarith.	Differen.	Logarit.	Sines.	
0	139173	1972036	1962257	9780	990268	60
1	139461	1969969	1960148	9820	990228	59
2	139749	1967905	1958044	9861	990187	58
3	140037	1965846	1955944	9903	990146	57
4	140325	1963792	1953848	9944	990105	56
5	140613	1961741	1951756	9985	990065	55
6	140901	1959695	1949669	10026	990024	54
7	141189	1957653	1947586	10068	989983	53
8	141477	1955616	1945507	10109	989942	52
9	141765	1953583	1943432	10151	989900	51
10	142053	1951554	1941361	10193	989859	50
11	142341	1949530	1939294	10234	989818	49
12	142629	1947508	1937232	10276	989776	48
13	142916	1945492	1935173	10318	989735	47
14	143205	1943479	1933119	10360	989693	46
15	143493	1941471	1931069	10402	989651	45
16	143780	1939467	1929022	10445	989610	44
17	144068	1937467	1926980	10487	989568	43
18	144356	1935471	1924941	10529	989526	42
19	144644	1933479	1922907	10572	989484	41
20	144932	1931491	1920876	10614	989442	40
21	145220	1929507	1918850	10657	989399	39
22	145507	1927527	1916828	10700	989357	38
23	145795	1925552	1914809	10743	989315	37
24	146083	1923580	1912794	10785	989272	36
25	146371	1921612	1910783	10828	989230	35
26	146659	1919648	1908776	10872	989187	34
27	146946	1917687	1906773	10915	989144	33
28	147234	1915731	1904773	10958	989102	32
29	147522	1913779	1902778	11001	989059	31
30	147809	1911831	1900786	11045	989016	30

Min.

Deg. 81

Deg. 8   +|−

m.	Sines.	Logarith.	Differen.	Logarit.	Sines.	
30	147809	1911831	1900786	11045	988016	30
31	148097	1909886	1898798	11088	988973	29
32	148385	1907946	1896814	11132	988930	28
33	148672	1906009	1890833	11176	988887	27
34	148960	1904076	1892857	11219	988843	26
35	149248	1902147	1890883	11263	988800	25
36	149535	1900221	1888914	11307	988756	24
37	149822	1898300	1886948	11351	988713	23
38	150110	1896382	1884987	11395	988669	22
39	150398	1894468	1883028	11440	988625	21
40	150686	1892558	1881074	11484	988582	20
41	150973	1890652	1879123	11528	988538	19
42	151261	1888749	1877176	11573	988494	18
43	151548	1886850	1875232	11617	988450	17
44	151836	1884954	1873292	11662	988406	16
45	152123	1883062	1871656	11707	988361	15
46	152411	1881174	1869423	11752	988317	14
47	152698	1879290	1867493	11797	988273	13
48	152986	1877409	1865567	11842	988228	12
49	153273	1875532	1863645	11887	988184	11
50	153561	1873658	1861726	11932	988139	10
51	153848	1871788	1859811	11977	988094	9
52	154136	1869922	1857899	12022	988050	8
53	154423	1868059	1855991	12068	988005	7
54	154710	1866199	1854086	12113	987960	6
55	154998	1864344	1852185	12159	987915	5
56	155285	1862491	1850287	12204	987870	4
57	155572	1860643	1848392	12250	987824	3
58	155860	1858797	1846501	12296	987779	2
59	156147	1856956	1844614	12342	987734	1
60	156434	1855117	1842729	12388	987688	0

Min.

Deg. 81

Deg. 9   +|−

m.	Sines.	Logarith.	Differen.	Logarit.	Sines.	
0	156434	1855117	1842729	12388	987688	60
1	156722	1853283	1840848	12434	987643	59
2	157009	1851451	1838971	12480	987597	58
3	157296	1849623	1837096	12527	987551	57
4	157584	1847798	1835225	12573	987506	56
5	157870	1845977	1833358	12620	987460	55
6	158158	1844159	1831493	12616	987414	54
7	158445	1842345	1829632	12713	987368	53
8	158732	1840534	1827775	12759	987322	52
9	159020	1838726	1825920	12806	987275	51
10	159307	1836922	1824069	12853	987229	50
11	159594	1835121	1822221	12900	987183	49
12	159881	1833324	1820376	12947	987136	48
13	160169	1831526	1818535	12994	987090	47
14	160455	1829738	1816697	13041	987043	46
15	160743	1827951	1814862	13089	986997	45
16	161030	1826166	1813030	13136	986950	44
17	161317	1824385	1811201	13184	986903	43
18	161604	1822607	1809376	13231	986856	42
19	161891	1820832	1807553	13279	986809	41
20	162178	1819061	1805734	13327	986762	40
21	162465	1817292	1803918	13375	986714	39
22	162752	1815527	1802104	13423	986667	38
23	163039	1813765	1800295	13471	986620	37
24	163326	1812007	1798488	13519	986572	36
25	163613	1810251	1796684	13567	986525	35
26	163900	1808499	1794883	13615	986477	34
27	164187	1806749	1793086	13664	986429	33
28	164474	1805003	1791291	13712	986381	32
29	164761	1803260	1789500	13761	986334	31
30	165048	1801521	1787711	13809	986286	30

Min.

Deg. 80

Deg. 9   +|−

m.	Sines.	Logarith.	Differen.	Logarit.	Sines.	
30	165048	1801521	1787711	13809	986286	30
31	165334	1799784	1785926	13858	986238	29
32	165621	1798050	1784143	13907	986189	28
33	165908	1796320	1782364	13956	986141	27
34	166195	1794592	1780587	14005	986093	26
35	166432	1792868	1778814	14054	986045	25
36	166769	1791146	1777043	14103	985996	24
37	167055	1789428	1775276	14152	985947	23
38	167342	1787713	1773511	14201	985899	22
39	167629	1786001	1771750	14251	985850	21
40	167916	1784291	1769991	14300	985801	20
41	166203	1782585	1768235	14350	985752	19
42	168489	1780882	1766482	14400	985703	18
43	168776	1779182	1764732	14449	985654	17
44	169063	1777484	1762985	14499	985605	16
45	169349	1775790	1761241	14549	985556	15
46	169636	1774098	1759499	14599	985507	14
47	169923	1772410	1757761	14649	985457	13
48	170209	1770724	1756025	14700	985408	12
49	170496	1769042	1754292	14750	985358	11
50	170783	1767362	1752562	14800	985309	10
51	171069	1765686	1750835	14851	985259	9
52	171356	1764012	1749111	14901	985209	8
53	171643	1762341	1747389	14952	985159	7
54	171929	1760673	1745670	15002	985109	6
55	172216	1759007	1743954	15053	985059	5
56	172502	1757345	1742241	15104	985009	4
57	172789	1755685	1740530	15155	984959	3
58	173075	1754028	1738822	15206	984909	2
59	173362	1752374	1737117	15257	984858	1
60	173648	1750723	1735415	15309	984808	0

Min.

Deg. 80

Deg. 10　　+|—

m.	Sines.	Logarith.	Differen.	Logarit.	Sines.	
0	173648	1750723	1735415	15309	984808	60
1	173935	1749075	1733715	15360	984757	59
2	174221	1747430	1732018	15411	984707	58
3	174507	1745787	1730324	15463	984656	57
4	174794	1744147	1728632	15515	984605	56
5	175080	1742519	1726943	15566	984554	55
6	175367	1740875	1925257	15618	984503	54
7	175653	1739244	1723574	15670	984452	53
8	175939	1737615	1721803	15722	984401	52
9	176226	1735989	1720215	15774	984350	51
10	176512	1734365	1718539	15826	984298	50
11	176798	1732744	1716866	15878	984247	49
12	177085	1731126	1715196	15931	984196	48
13	177371	1729511	1713528	15983	984144	47
14	177657	1727898	1711863	16035	984092	46
15	177943	1726288	1710200	16088	984041	45
16	178230	1724681	1708540	16141	983989	44
17	178516	1723076	1706883	16193	983937	43
18	178802	1721474	1705228	16246	983885	42
19	179088	1719875	1703576	16299	983833	41
20	179375	1718278	1701926	16352	983781	40
21	179661	1716684	1700279	16405	983729	39
22	179947	1715093	1698634	16458	983676	38
23	180233	1713504	1696992	16512	983624	37
24	180519	1711918	1695353	16565	983571	36
25	180805	1710334	1693716	16618	983519	35
26	181091	1708753	1692081	16672	983466	34
27	181377	1707175	1690449	16725	983414	33
28	181663	1705599	1688819	16779	983360	32
29	181949	1704025	1687192	16833	983308	31
30	182235	1702454	1685568	16887	983255	30

Min.

Deg. 79

Deg. 10 $+|-$

m.	Sines.	Logarith.	Differen.	Logarit.	Sines.	
30	182235	1702455	1685568	16887	983255	30
31	182521	1700887	1683946	16941	983202	29
32	182807	1699321	1682326	16994	983149	28
33	183093	1697758	1680709	17049	983096	27
34	183379	1696197	1679094	17103	983042	26
35	183665	1694639	1677482	17157	982989	25
36	183951	1693084	1675872	17212	982935	24
37	184237	1691530	1674264	17266	982882	23
38	104523	1689980	1672659	17321	982828	22
39	184809	1688432	1671056	17376	982774	21
40	185095	1686886	1669456	17430	982721	20
41	185381	1685343	1667858	17485	982667	19
42	185667	1683802	1666262	17540	982613	18
43	185952	1682264	1664669	17595	982559	17
44	186238	1680728	1663078	17650	982505	16
45	186524	1679195	1661489	17705	982450	15
46	186810	1677664	1659903	17761	982396	14
47	187096	1676135	1658319	17816	982342	13
48	187381	1674609	1656737	17871	982287	12
49	187667	1673085	1655158	17927	982223	11
50	187953	1671564	1653581	17983	982178	10
51	188238	1670045	1652007	18038	982123	9
52	188524	1668529	1650434	18094	982069	8
53	188810	1667014	1648864	18150	982014	7
54	189095	1665503	1647297	18206	981959	6
55	189381	1663993	1645731	18262	981904	5
56	189667	1662486	1644168	18318	981849	4
57	189952	1660982	1642607	18374	981793	3
58	190238	1659479	1641049	18431	981738	2
59	190523	1657979	1639492	18487	981683	1
60	190809	1656482	1637938	18544	981627	0

Min.

Deg. 79

Deg. 11   +|—

m.	Sines.	Logarith.	Differen.	Logarit.	Sines.	
0	190809	1656481	1637938	18544	981627	60
1	191094	1654980	1636386	18600	981572	59
2	191380	1653493	1634836	18657	981516	58
3	191665	1652003	1633289	18714	981460	57
4	191951	1650514	1631744	18770	981404	56
5	192236	1649028	1630201	18827	831349	55
6	192522	1647594	1628660	18884	981293	54
7	192807	1646063	1627121	18942	981237	53
8	193093	1644584	1625585	18999	981185	52
9	193378	1643107	1624051	19056	981124	51
10	193664	1641632	1622518	19113	981068	50
11	193949	1640160	1620989	19171	981012	49
12	194234	1638689	1619461	19228	980955	48
13	194520	1637222	1617935	19286	980899	47
14	194805	1635756	1616412	19344	980842	46
15	195090	1634292	1614891	19402	980785	45
16	195376	1632831	1613372	19459	980728	44
17	195661	1631372	1611854	19517	980672	43
18	165946	1629915	1610339	16576	980615	42
19	196231	1628460	1608827	19634	980558	41
20	196517	1627008	1607316	19692	980501	40
21	196802	1625550	1605807	19750	980443	39
22	197087	1624109	1604301	19809	980386	38
23	197372	1622664	1602796	19867	980329	37
24	197657	1601220	1601294	19926	980271	36
25	197942	1619778	1599794	19984	980214	35
26	198228	1618339	1598295	20043	980156	34
27	198513	1616902	1596799	20102	980098	33
28	198798	1615466	1595305	20161	980041	32
29	199083	1614034	1593813	20220	979983	31
30	199368	1612603	1592323	20279	979925	30

Min.

Deg. 78

Deg. 11 +|−

m.	Sines.	Logarith.	Differen.	Logarit.	Sines.	
30	199368	1612603	1592323	20279	979925	30
31	199653	1611174	1590835	20339	979867	29
32	199938	1609748	1589350	20398	979809	28
33	200223	1608323	1587866	20457	979750	27
34	200508	1606901	1586384	20517	979692	26
35	200793	1605481	1584904	20576	979634	25
36	201078	1604062	1583426	20636	979575	24
37	201363	1602646	1581950	20696	979517	23
38	201648	1601232	1580476	20756	979458	22
39	201933	1599820	1579005	20816	979399	21
40	202218	1598411	1477534	20876	979341	20
41	202503	1597003	1576067	20936	979282	19
42	202787	1595597	1574601	20996	979223	18
43	203072	1594194	1573137	21056	979164	17
44	203357	1592792	1571675	21117	979105	16
45	203642	1591393	1570215	21177	979046	15
46	203927	1589995	1568757	21238	978986	14
47	204211	1588600	1567301	21298	978927	13
48	204496	1587206	1565846	21359	978867	12
49	204781	1585815	1564395	21420	978808	11
50	205066	1584425	1562944	21481	978749	10
51	205350	1583037	1561493	21542	978689	9
52	205635	1581652	1560049	21603	978629	8
53	205920	1580269	1558604	21664	978569	7
54	206204	1578887	1557162	21725	978509	6
55	206489	1577508	1555721	21787	978449	5
56	206774	1576130	1554282	21848	978389	4
57	207058	1574750	1552840	21910	978319	3
58	207343	1573382	1551411	11971	978268	2
59	207627	1972011	1449978	22033	978208	1
60	207912	1570641	1548547	22095	978148	0

Min.

Deg. 78

Deg. 12   +|−

m.	Sines.	Logarith.	Differen.	Logarit.	Sines.	
0	207912	1570641	1548547	22095	978148	60
1	208196	1569274	1547117	22157	978087	59
2	028481	1567908	1545690	22219	978026	58
3	208765	1566544	1544264	22281	977966	57
4	209050	1565183	1542840	22343	977905	56
5	209334	1563823	1541418	22405	977844	55
6	209618	1562465	1539998	22467	977783	54
7	209903	1561109	1538580	22530	977722	53
8	210187	1559755	1537163	22592	977661	52
9	210472	1558403	1535748	22655	977600	51
10	210756	1557053	1534336	22717	977539	50
11	211040	1555705	1532925	22780	977477	49
12	211325	1554358	1531515	22843	977416	48
13	211609	1553014	1530108	22906	977354	47
14	211893	1551671	1528703	22969	977293	46
15	212178	1550331	1527299	23032	977231	45
16	212462	1548992	1525897	23095	977169	44
17	212746	1547655	1524497	23158	977107	43
18	213030	1546320	1523098	23222	977046	42
19	213315	1544987	1521701	23285	976984	41
20	213599	1543655	1520306	23349	976921	40
21	213883	1542326	1518913	23413	976859	39
22	214167	1540998	1517522	23476	976797	38
23	214451	1539672	1516132	23540	976735	37
24	214735	1538348	1514744	23604	976672	36
25	215019	1537026	1513358	23668	976610	35
26	215303	1535706	1511974	23732	976547	34
27	215588	1534387	1510591	23796	976484	33
28	215872	1533071	1509210	23861	976422	32
29	216156	1531756	1507831	23925	976359	31
30	216440	1530443	1506453	23989	976297	30

Min.

Deg. 77

Deg. 12 + | −

m.	Sines.	Logarith.	Differen.	Logarit.	Sines.	
30	216440	1530443	1506453	23989	976296	30
31	216724	1529132	1505078	24054	976233	29
32	217008	1527823	1503704	24119	976170	28
33	217292	1526515	1502332	24183	976107	27
34	217575	1525209	1500961	24248	976043	26
35	217859	1523905	1499592	24313	975980	25
36	218143	1522603	1498225	24378	975917	24
37	218427	1521302	1496859	24443	975853	23
38	218711	1520004	1495495	24508	975790	22
39	218995	1518707	1494133	24573	975726	21
40	219279	1517412	1492773	24639	975662	20
41	219562	1516118	1491414	24704	975598	19
42	219846	1514827	1490057	24770	975535	18
43	220130	1513537	1488701	24835	975470	17
44	220414	1512248	1487347	24901	975406	16
45	220697	1510962	1485995	24967	975342	15
46	220981	1509677	1484645	25033	975278	14
47	221265	1508394	1483206	25099	975214	13
48	221548	1507113	1481940	25165	975149	12
49	221832	1505834	1480603	25231	975085	11
50	222116	1504556	1479259	25297	975020	10
51	222400	1503280	1477917	25363	974956	9
52	222683	1502006	1476576	25430	974891	8
53	222967	1500733	1475237	25496	974826	7
54	223250	1499462	1473899	25563	974761	6
55	223534	1498193	1472563	25629	974696	5
56	223817	1496925	1471229	25696	974631	4
57	224101	1495659	1469896	25763	974566	3
58	224384	1494395	1468565	25830	974501	2
59	224668	1493132	1467235	25897	974435	1
60	224951	1491872	1465908	25964	974370	0

Min.

Deg. 77

Deg. 13   + | −

m.	Sines.	Logarith.	Differen.	Logarit.	Sines.	
0	224951	1491872	1465907	25964	974370	60
1	225234	1490612	1464581	26031	974304	59
2	225518	1489355	1463256	26097	974239	58
3	225801	1488099	1461933	26166	974173	57
4	226085	1486845	1460612	26234	974108	56
5	226368	1485593	1459291	26301	974042	55
6	226651	1484341	1457973	26369	973976	54
7	226935	1483093	1456656	26436	973910	53
8	227217	1481845	1455341	26504	973843	52
9	227501	1480599	1454027	26572	973778	51
10	227784	1479355	1452715	26640	973711	50
11	228068	1478113	1451405	26708	973645	49
12	228351	1476872	1450095	26776	973579	48
13	228634	1475632	1448788	26845	973512	47
14	228917	1474395	1447482	26913	973446	46
15	229200	1473158	1446177	26981	973379	45
16	229483	1471924	1444874	27050	973313	44
17	229767	1470691	1443572	27118	973246	43
18	230050	1469459	1442274	27187	973179	42
19	230332	1468230	1440974	27256	973112	41
20	230610	1467001	1439677	27325	973045	40
21	230891	1465775	1438381	27394	972978	39
22	231181	1464550	1437087	27463	972910	38
23	231465	1463326	1435794	27532	972843	37
24	231748	1462104	1434503	27601	972776	36
25	232030	1460884	1433213	27671	972708	35
26	232314	1459665	1431925	27740	972640	34
27	232597	1458448	1430638	27810	972573	33
28	232880	1457233	1429353	27879	972506	32
29	233162	1456019	1428070	27949	972438	31
30	233445	1454807	1426788	28019	972310	30

Min.

Deg. 76

Deg. 13   + | −

m.	Sines.	Logarith.	Differen.	Logarit.	Sines.	
30	233445	1454807	1426788	28019	972370	30
31	233728	1453596	1425507	28089	972302	29
32	234011	1452387	1424228	28159	972234	28
33	234294	1451179	1422950	28229	972166	27
34	234577	1449973	1421674	38299	972098	26
35	234859	1448768	1420399	28369	972029	25
36	235142	1447565	1419125	28439	971961	24
37	235425	1446363	1417853	28510	971893	23
38	235707	1445163	1416583	28580	971824	22
39	235990	1443965	1415313	28651	971755	21
40	236273	1442767	1414046	28722	971687	20
41	236555	1441572	1412779	28792	971618	19
42	236838	1440378	1411514	28863	971549	18
43	237121	1439185	1410251	28934	971480	17
44	237403	1437994	1408989	29005	971411	16
45	237686	1436805	1407728	29076	971342	15
46	237968	1435616	1406469	29148	971272	14
47	238251	1434430	1405211	29219	971204	13
48	230533	1433245	1403955	29290	971134	12
49	238816	1432862	1402700	29362	971065	11
50	239098	1430080	1401446	29433	970995	10
51	239381	1429699	1400194	29505	970926	9
52	239663	1428520	1398943	29577	970856	8
53	239946	1327342	1397693	29649	970786	7
54	240228	1426166	1396445	29721	970716	6
55	240510	1424991	1395199	29792	970647	5
56	240793	1423818	1393953	29865	970577	4
57	241075	1422646	1392709	29937	970506	3
58	241357	1421476	1391467	30009	970436	2
59	241640	1420307	1390225	30082	970366	1
60	241922	1419140	1388985	30154	970296	0

Min.

Deg. 76

Deg. 14   +|−

m.	Sines.	Logarith.	Differen.	Logarit.	Sines.	
0	241922	1419140	1388985	30154	970296	60
1	242204	1417974	1387747	30227	970225	59
2	242486	1416809	1386509	30300	970155	58
3	242768	1415645	1385274	30372	970084	57
4	243051	1414484	1384039	30445	970013	56
5	243333	1413324	1382806	30518	969943	55
6	243615	1412165	1381574	30591	969872	54
7	243897	1411008	1380344	30664	969801	53
8	244179	1409852	1379115	30737	969730	52
9	244461	1408698	1377887	30811	969659	51
10	244743	1407545	1376661	30884	969588	50
11	245025	1406393	1375435	30958	969517	49
12	245307	1405243	1374212	31031	969445	48
13	245589	1404094	1372989	31105	969374	47
14	245871	1402946	1371768	31179	969302	46
15	246153	1401800	1370548	31252	969231	45
16	246435	1400656	1369329	31326	969159	44
17	246717	1399512	1368112	31400	969088	43
18	246999	1398370	1366896	31474	969016	42
19	247281	1397230	1365681	31549	968944	41
20	247563	1396091	1364468	31623	968872	40
21	247845	1394953	1363256	31697	968800	39
22	248126	1393817	1362045	31772	968728	38
23	248408	1392682	1360835	31846	968655	37
24	248690	1391548	1359627	31921	968583	36
25	248972	1390416	1358420	31996	968511	35
26	249253	1389285	1357214	32070	968438	34
27	249535	1388155	1356010	32145	968306	33
28	249817	1387027	1354807	32220	968293	32
29	250098	1385900	1353605	32295	968225	31
30	250380	1384775	1352404	32371	968148	30

Min.

Deg. 75

Deg. 14   +|−

m.	Sines.	Logarith.	Differen.	Logarit.	Sines.	
30	250380	1384775	1352404	32371	968148	30
31	250663	1383651	1351205	32446	968075	29
32	250943	1382528	1350007	32521	968002	28
33	251225	1381407	1348810	32597	967929	27
34	251506	1380286	1347614	32672	967856	26
35	251788	1379168	1346420	32748	967782	25
36	252069	1378050	1345227	32824	967709	24
37	252351	1376934	1344035	32899	967636	23
38	252632	1375819	1342844	32975	967562	22
39	252914	1374706	1341655	33051	967489	21
40	253195	1373594	1340466	33127	967415	20
41	253477	1372483	1339280	33204	967341	19
42	253758	1371374	1338094	33280	967268	18
43	254039	1370266	1336910	33356	967194	17
44	254321	1369159	1335726	33433	967120	16
45	254602	1368053	1334544	33509	967046	15
46	254883	1366949	1333363	33586	966972	14
47	255164	1365846	1332184	33663	966898	13
48	255446	1364744	1331005	33739	966823	12
49	255727	1363644	1329828	33816	966749	11
50	256008	1362545	1328652	33893	966675	10
51	256289	1361447	1327477	33970	966600	9
52	256571	1360351	1326303	34048	966525	8
53	256852	1359256	1325131	34125	966451	7
54	257133	1358152	1323960	34202	966376	6
55	257414	1357069	1322790	34280	966301	5
56	257695	1355978	1321621	34357	966226	4
57	257976	1354888	1320453	34435	966151	3
58	258257	1353799	1319287	34513	966076	2
59	258538	1352711	1318121	34590	966001	1
60	258819	1351625	1316557	34668	965926	0

Min.

Deg. 75

Deg. 15   +|−

m.	Sines.	Logarith.	Differen.	Logarit.	Sines.	
0	258819	1351625	1316557	34668	965927	60
1	259100	1350541	1315794	34746	965850	59
2	259381	1349457	1314633	34824	965775	58
3	259662	1348375	1313472	34903	964700	57
4	259943	1347293	1312313	34980	964624	56
5	269224	1346213	1311154	35059	965548	55
6	260504	1345135	1309997	35137	965473	54
7	260785	1344057	1308841	35216	965396	53
8	261066	1342981	1307686	35295	965321	52
9	261347	1341906	1306533	35373	965245	51
10	261628	1340832	1305380	35452	965169	50
11	261908	1339760	1304229	35531	965093	49
12	262189	1338688	1303078	35610	965016	48
13	262470	1337618	1301929	35689	964940	47
14	262750	1336549	1300781	35768	964864	46
15	263031	1335482	1299634	35848	964787	45
16	263312	1334415	1298488	35947	964711	44
17	263592	1333350	1297344	36006	964634	43
18	263873	1332286	1296200	36086	664557	42
19	264154	1331224	1295058	36165	964481	41
20	264434	1330162	1293917	36245	964403	40
21	264715	1329102	1292777	36325	964327	39
22	264995	1328043	1291638	36405	964250	38
23	265276	1326985	1290500	36485	964173	37
24	265556	1325929	1289364	36565	964095	36
25	265837	1324873	1288228	36645	964018	35
26	266117	1323819	1287094	36725	963941	34
27	266397	1322766	1285960	36806	963863	33
28	266678	1321714	1284828	36886	963786	32
29	266958	1320663	1283696	36967	963708	31
30	267238	1319614	1282566	37047	963630	30

Min.

Deg. 74

Deg. 15 + | −

m.	Sines.	Logarith.	Differen.	Logarit.	Sines.	
30	267238	1319613	1282566	37047	963630	30
31	267519	1318565	1281437	37128	963553	29
32	267799	1317518	1280309	37209	963475	28
33	268079	1316472	1279182	37290	963397	27
34	268359	1315427	1278056	37371	963319	26
35	268640	1314383	1276932	37452	963241	25
36	268920	1313341	1275808	37533	963162	24
37	269200	1312300	1274686	37614	963084	23
38	269480	1311259	1273564	37696	963016	22
39	269760	1310221	1272444	37777	962927	21
40	270040	1309183	1271325	37859	962849	20
41	270320	1308146	1270206	37940	962772	19
42	270600	1307111	1269089	38022	962692	18
43	270880	1306077	1267973	38104	962613	17
44	271160	1305044	1266858	38186	962534	16
45	271440	1304012	1265744	38268	962455	15
46	271720	1302981	1264631	38350	962376	14
47	272000	1301951	1263519	38432	962297	13
48	272280	1300922	1262409	38514	962218	12
49	272560	1299895	1261299	38597	962139	11
50	272840	1298869	1260190	38679	962059	10
51	273120	1297844	1259082	38762	961980	9
52	273400	1296820	1257976	38844	961900	8
53	273679	1295797	1256870	38927	961821	7
54	273959	1294775	1255766	39010	961741	6
55	274239	1293754	1254662	39092	961662	5
56	274519	1292735	1253560	39176	961582	4
57	274798	1291717	1252458	39259	961502	3
58	275078	1290699	1251358	39342	961422	2
59	275350	1289683	1250259	39425	961342	1
60	275637	1288668	1249160	39509	961262	0

Min.

G **Deg. 74**

Deg. 16 +|−

m.	Sines.	Logarith.	Differen.	Logarit.	Sines.	
0	275637	1288669	1249160	39509	961262	60
1	275917	1287655	1248063	39592	961181	59
2	276196	1286642	1246967	39676	961101	58
3	276476	1285631	1245871	39759	961021	57
4	276756	1284620	1244777	39843	960940	56
5	277035	1283610	1243684	39927	960860	55
6	277315	1282602	1242591	40010	960790	54
7	277594	1281595	1241500	40091	960698	53
8	277873	1280589	1240410	40179	960618	52
9	278153	1279583	1239320	40263	960537	51
10	278432	1278579	1238232	40347	960456	50
11	278712	1277577	1237145	40432	960375	49
12	278991	1276575	1236059	40516	960294	48
13	279270	1275574	1234973	40601	960213	47
14	279550	1274574	1233889	40685	960131	46
15	279830	1273576	1232806	40770	960050	45
16	280108	1272578	1231723	40855	959968	44
17	280387	1271582	1230642	40940	959887	43
18	280667	1270587	1229562	41025	959805	42
19	280946	1269592	1228483	41110	959724	41
20	282225	1268599	1227404	41195	959642	40
21	281504	1267607	1226327	41280	959560	39
22	281783	1266617	1225251	41366	959478	38
23	282062	1265627	1224175	41452	959396	37
24	282341	1264638	1223101	41537	959314	36
25	282624	1263650	1222027	41623	959232	35
26	282899	1262663	1228955	41708	959149	34
27	283178	1261678	1219783	41794	959067	33
28	283457	1260693	1218813	41880	958985	32
29	283736	1259709	1217743	41966	958902	31
30	284015	1258727	1216675	42052	958820	30

Min.

Deg. 73

Deg. 16 + | −

m.	Sines.	Logarith.	Differen.	Logarit.	Sines.	
30	284015	1258727	1216675	42052	958820	30
31	284294	1257745	1215607	42138	958737	29
32	284573	1256765	1214540	42225	958654	28
33	284852	1255785	1213474	42311	958572	27
34	285131	1254807	1212409	42397	958489	26
35	285410	1253830	1211345	42484	958406	25
36	285688	1252853	1210282	42571	958323	24
37	285967	1251878	1209220	42658	958239	23
38	286246	1250904	1208159	42744	958156	22
39	286525	1249930	1207099	42831	958073	21
40	286803	1248958	1206040	42918	957990	20
41	287082	1247987	1204982	43005	957906	19
42	287360	1247017	1203925	43093	957822	18
43	287639	1246048	1202868	43180	957739	17
44	287918	1245080	1201813	43267	957655	16
45	288196	1244113	1200758	43355	957571	15
46	288475	1243147	1199705	43442	957487	14
47	288753	1242182	1198652	43530	957404	13
48	289032	1241218	1197600	43618	957320	12
49	289310	1240255	1196549	43706	957235	11
50	289589	1239293	1195500	43794	957151	10
51	289867	1238332	1194451	43882	957067	9
52	290146	1237372	1193402	43970	956983	8
53	290424	1236413	1192355	44058	956898	7
54	290702	1235455	1191309	44146	956814	6
55	290981	2234498	1190264	44232	956729	5
56	291159	1233542	1189219	44323	956644	4
57	291537	1232588	1188176	44412	956560	3
58	291815	1231634	1187133	44501	956475	2
59	292093	1230681	1186091	44590	956390	1
60	292372	1229728	1185050	44679	956305	0

Min.

G 2 **Deg. 73**

Deg. 17   +|−

m.	Sines.	Logarith.	Differen.	Logarit.	Sines.	
0	292372	1229729	1185050	44678	956305	60
1	292650	1228778	1184010	44767	956220	59
2	292928	1227828	1182971	44858	956134	58
3	293206	1226879	1181933	44945	956049	57
4	293484	1225931	1180896	45065	955964	56
5	293762	1224984	1179859	45124	955878	55
6	294040	1224038	1178824	45214	955793	54
7	294318	1223093	1177789	45303	955707	53
8	294596	1222149	1176756	45393	955622	52
9	294874	1221206	1175723	45482	955536	51
10	295152	1221206	1174691	45572	955450	50
11	295430	1219322	1173660	45662	955364	49
12	295708	1218381	1172619	45752	955278	48
13	295986	1217443	1171600	45842	955192	47
14	296263	1216504	1170572	45932	955106	46
15	296542	1215567	1169544	46023	955020	45
16	296819	1214631	1168517	46113	954934	44
17	297097	1213695	1167491	46104	954847	43
18	297375	1212761	1166466	46294	954762	42
19	297653	1211828	1165442	46385	954674	41
20	297930	1210895	1164419	46475	954588	40
21	298208	1209964	1163397	46566	954501	39
22	298486	1209033	1162376	46657	954414	38
23	298763	1208104	1161353	46748	954327	37
24	299041	1207175	1160335	46839	954240	36
25	299318	1206247	1159316	46930	954153	35
26	299596	1205320	1158298	47022	954066	34
27	299873	1204394	1157281	47113	953979	33
28	300251	1203470	1156165	47105	953892	32
29	300428	1202546	1155249	47296	953804	31
30	300706	1201621	1154234	47388	953717	30

Min.

Deg. 72

Deg. 17   +|—

m.	Sines.	Logarith.	Differen.	Logarit.	Sines.	
30	300706	1201622	1154234	47388	953717	30
31	300983	1200700	1153220	47480	953629	29
32	301261	1199779	1152207	47572	953541	28
33	301538	1198859	1151195	47664	953454	27
34	301815	1197940	1150183	47756	953366	26
35	302093	1197021	1149173	47848	953279	25
36	302370	1196104	1148163	47940	953191	24
37	302647	1195187	1147154	48033	953103	23
38	302924	1194272	1146146	48125	953015	22
39	303202	1193357	1145139	48218	952926	21
40	303478	1192443	1144135	48310	952838	20
41	303756	1191530	1143127	48403	952750	19
42	304033	1190618	1142123	48496	952661	18
43	304310	1189707	1141119	48589	952573	17
44	304587	1188797	1140116	48692	952481	16
45	304864	1187888	1139113	48775	952396	15
46	305141	1186980	1138112	48868	952307	14
47	305418	1186072	1137111	48961	952218	13
48	305695	1185166	1136111	49054	952129	12
49	305972	1184260	1135112	49148	952040	11
50	306249	1183356	1134114	49241	951951	10
51	306526	1182452	1133117	49335	951862	9
52	306803	1181549	1132121	49429	951773	8
53	307080	1180647	1131125	49522	951684	7
54	307357	1179746	1130130	49616	951594	6
55	307633	1178846	1129136	49710	951505	5
56	307910	1177947	1128142	49804	951415	4
57	308187	1177040	1127150	49899	951326	3
58	308464	1176151	1126158	49993	951236	2
59	308740	1175254	1125167	50087	651146	1
60	309017	1174359	1124177	50181	951056	0

Min.

G 3   **Deg. 72**

Deg. 18   +|−

m.	Sines.	Logarith.	Differen.	Logarit.	Sines.	
0	309017	1174359	1124177	50182	951056	60
1	309294	1173464	1123187	50176	950967	59
2	309570	1172579	1122199	50371	950879	58
3	309847	1171677	1121211	50466	950787	57
4	310123	1170785	1120224	50561	950696	56
5	310400	1169893	1119238	50656	950606	55
6	310676	1169003	1118252	50751	950516	54
7	310953	1168113	1117268	50846	950425	53
8	311229	1167225	1116284	50941	950335	52
9	311506	1166337	1115301	51036	950244	51
10	311782	1165450	1114318	51132	950154	50
11	312059	1164564	1113337	51227	950063	49
12	312335	1163679	1112356	51323	949972	48
13	312611	1162794	1111376	51418	949881	47
14	312887	1161911	1110397	51514	949790	46
15	313164	1161028	1109418	51610	949699	45
16	313440	1160143	1108440	51706	949608	44
17	313716	1159261	1107464	51802	949517	43
18	313992	1158386	1106488	51898	949425	42
19	314269	1157507	1105517	51994	949334	41
20	314545	1156628	1124538	52091	949243	40
21	314821	1155751	1103564	52187	949151	39
22	315097	1154875	1102591	52284	949060	38
23	315373	1153999	1101618	52380	948968	37
24	315649	1153124	1100647	52477	948876	36
25	315925	1254250	1099676	52574	948784	35
26	316201	1151377	1098706	52670	948692	34
27	316473	1150504	1097737	52768	948600	33
28	316753	1149633	1096768	52865	948508	32
29	317029	1148762	1095800	52962	948416	31
30	317305	1147893	1094633	53059	948324	30

Min.

Deg. 71

Deg. 18   + | —

m.	Sines.	Logarith.	Differen.	Logarit.	Sines.	
30	317305	1147893	1094833	53059	948314	30
31	317580	1147024	1093867	53157	948231	29
32	317856	1146156	1092901	53254	948139	28
33	318132	1145288	1091936	53352	948046	27
34	318408	1144422	1090972	53450	947954	26
35	318684	1143556	1090009	53547	947861	25
36	318959	1142691	1089046	53645	947768	24
37	319235	1141827	1088084	53743	947676	23
38	319511	1140964	1087123	53841	947583	22
39	319786	1140102	1086163	53939	947490	21
40	320062	1139241	1085203	54037	947397	20
41	320337	1138380	1084244	54136	947303	19
42	320613	1137520	1083386	54234	947210	18
43	320888	1136661	1082329	54332	947117	17
44	321164	1135803	1081372	54431	947024	16
45	321439	1134946	1080416	54530	946930	15
46	321715	1134089	1079460	54629	946837	14
47	321990	1133233	1078506	54727	946743	13
48	322266	1132378	1077552	54826	946649	12
49	322541	1131524	1076599	54926	946555	11
50	322816	1130671	1075646	55025	946462	10
51	323092	1129819	1074694	55124	946368	9
52	323367	1128967	1073743	55224	946174	8
53	323642	1128116	1072793	55323	946180	7
54	323917	1127266	1071844	55423	946085	6
55	324193	1126417	1070895	55522	945991	5
56	324468	1125569	1069947	55612	945897	4
57	324743	1124721	1068999	55722	945802	3
58	325018	1123874	1068053	55822	945705	2
59	325293	1123028	1067107	55922	945613	1
60	325568	1122183	1066161	56022	945115	0

Min.

G 4   **Deg. 71**

Deg. 19   +|−

m.	Sines.	Logarith.	Differen.	Logarit.	Sines.	
0	325568	1122183	1066161	56022	945519	60
1	325843	1121339	1065217	56122	945424	59
2	316118	1120495	1064273	56222	945329	58
3	326393	1119652	1063330	56323	945234	57
4	326668	1118810	1062387	56423	945139	56
5	326943	1117969	1061445	56524	945041	55
6	327218	1117129	1060504	56624	944949	54
7	327493	1116189	1059564	56725	944854	53
8	327767	1115450	1058624	56826	944758	52
9	328042	1114612	1057685	56927	944663	51
10	328317	1113775	1056747	57028	944568	50
11	328592	1112938	1055809	57129	944472	49
12	329866	1112102	1054872	57230	944376	48
13	329141	1111267	1053936	57332	944281	47
14	329416	1110433	1053000	57433	944185	46
15	329691	1109600	1052065	57534	944089	45
16	329965	1108767	1051131	57636	943993	44
17	330240	1107936	1050198	57738	943897	43
18	330514	1107105	1049265	57840	943801	42
19	330789	1106274	1048333	57942	943705	41
20	331063	1105445	1047401	58044	943608	40
21	331338	1104616	1046470	58146	943512	39
22	331612	1103788	1045540	58248	943416	38
23	331887	1102961	1044611	58350	943319	37
24	332161	1102135	1043682	58453	943223	36
25	332455	1101309	1042754	58555	943126	35
26	332710	1100484	1041816	58658	943029	34
27	332984	1099660	1040899	58761	942932	33
28	333258	1098837	1039973	58863	942836	32
29	333533	1098024	1039048	58966	942739	31
30	333807	1097192	103123	59069	942641	30

Min.

Deg. 70

Deg. 19   +|−

m.	Sines.	Logarith.	Differen.	Logarit.	Sines.	
30	333807	1097192	1038123	59069	942641	30
31	334081	1096371	1037199	59172	942544	29
32	334355	1095551	1036276	59275	642447	28
33	334629	1094731	1035353	59378	942350	27
34	334903	1093912	1034430	59482	942252	26
35	335178	1093094	1033509	59585	942155	25
36	335452	1092277	1032588	59689	942057	24
37	335726	1091461	1031668	59792	941960	23
38	336000	1090645	1030749	59896	941862	22
39	336274	1089830	1029830	60000	941764	21
40	336547	1089016	1028911	60104	941666	20
41	336821	1088202	1027994	60208	942568	19
42	337095	1087389	1027077	60312	941470	18
43	337369	1086577	1026161	60416	941372	17
44	337642	1085766	1025245	60520	941274	16
45	337917	1084955	1024330	60625	941176	15
46	338190	1084146	1023416	60730	941078	14
47	338464	1083337	1022502	60834	940979	13
48	338738	1082528	1021589	60939	943882	12
49	339011	1081721	1820677	61044	940782	11
50	339285	1080914	1019765	61148	940684	10
51	339559	1080107	1018854	61253	940585	9
52	339832	1079302	1017944	61358	940486	8
53	340106	1073497	1017034	61463	940387	7
54	340379	1077693	1016125	61569	940288	6
55	340653	1076890	1015216	61674	940189	5
56	340926	1076088	1014308	61779	940090	4
57	341200	1075286	1013401	61885	939991	3
58	331473	1074185	1012249	61991	939891	2
59	341747	1073685	1011588	62097	939792	1
60	342020	1072885	1010683	62202	939693	0

Min.

G 5   **Deg. 70**

Deg. 20   +|—

m.	Sines.	Logarith.	Differen.	Logarit.	Sines.	
0	342020	1072883	1010683	62202	939693	60
1	342293	1072086	1009778	62308	939593	59
2	342567	1071283	1008874	62414	939493	58
3	342840	1070492	1007971	62520	939394	57
4	343113	1069694	1007068	62627	639294	56
5	343386	1068893	1006165	62733	939194	55
6	343660	1068103	1005264	62839	939094	54
7	343933	1067308	1004363	62946	938994	53
8	334206	1066515	1003462	63052	938894	52
9	344479	1065722	1002562	63159	938794	51
10	344752	1064929	1001663	63266	938694	50
11	345025	1064137	1000765	63373	938593	49
12	345298	1063346	999867	63480	938493	48
13	345571	1062556	998969	63587	938392	47
14	345844	1061767	998072	63694	938292	46
15	346117	1060978	997176	63801	938191	45
16	346390	1060190	996281	63909	938091	44
17	346663	1059402	995386	64016	937990	43
18	346936	1058616	994492	64124	937889	42
19	347108	1057830	993598	64231	937788	41
20	347481	1057044	992705	64339	937687	40
21	347754	1056260	991813	64447	937586	39
22	348027	1055476	990921	64555	937485	38
23	348109	1054693	990030	64662	937383	37
24	348572	1053910	989140	64771	937282	36
25	348845	1053129	988250	64879	937181	35
26	349117	1052347	987360	64987	937079	34
27	349390	1051567	986471	65096	936977	33
28	349662	1050787	985583	65204	936876	32
29	349934	1050008	984694	65313	936774	31
30	350207	1049219	983808	65422	936672	30

Min.

Deg. 69

Deg. 20 +|−

m.	Sines.	Logarith.	Differen.	Logarit.	Sines.	
30	350207	1049219	983808	65422	936672	30
31	350480	1048452	982921	65531	936570	29
32	350752	1047674	982035	65640	636468	28
33	351025	1046898	981149	65749	936360	27
34	351297	1046122	980265	65858	936264	26
35	351569	1845348	979381	65967	936162	25
36	351842	1044573	978497	66076	936059	24
37	352214	1043800	977614	66186	935957	23
38	352386	1043027	976732	66295	935855	22
39	352658	1042255	975850	66405	935752	21
40	352931	1041484	974969	66514	935649	20
41	353203	1040713	974089	66624	935547	19
42	353475	1039943	973209	66734	935444	18
43	353747	1039173	972330	66844	935341	17
44	354029	1038405	971451	66954	935238	16
45	354291	1037637	970573	67064	935135	15
46	354563	1036869	969695	67174	935032	14
47	354835	1036102	968818	67284	934929	13
48	355107	1035336	967941	67395	934826	12
49	355379	1034571	967065	67506	934722	11
50	355561	1033806	966189	67616	934619	10
51	355923	1033041	965314	67727	934515	9
52	356194	1032278	964440	67838	934412	8
53	356466	1031515	963566	67949	934308	7
54	356738	1030753	962693	68060	934204	6
55	357010	1029992	961820	68171	934101	5
56	357281	1029231	960948	68282	933997	4
57	357553	1828471	960077	68394	933893	3
58	357825	1027711	959206	68505	933789	2
59	358096	1026953	959336	68617	933685	1
60	358368	1026195	957466	68728	933580	0

Min.

Deg. 69

Deg. 21   +|−

m.	Sines.	Logarith.	Differen.	Logarit.	Sines.	
0	358368	1026195	957466	68728	933580	60
1	358639	1025437	956597	68840	933476	59
2	358911	1024680	955729	68952	933372	58
3	359182	1023924	954861	96064	933267	57
4	359454	1023169	953993	96176	933163	56
5	359725	1022414	953126	96288	933058	55
6	359997	1021660	952260	69400	932953	54
7	360268	1020906	951394	69512	932849	53
8	360539	1020153	950529	69625	932714	52
9	360811	1019401	949664	69737	932639	51
10	361082	1018650	948800	69849	932534	50
11	361353	1017899	947937	69962	932429	49
12	361624	1017148	947074	70075	392324	48
13	361896	1016399	946211	70188	392219	47
14	362167	1015650	645349	70301	392113	46
15	362438	1014901	944488	70414	932008	45
16	362709	1014154	943627	70527	931902	44
17	362980	1013407	942766	70640	931797	43
18	363251	1012660	941907	70754	931691	42
19	363522	1011914	941047	70867	931586	41
20	363793	1011169	940189	70981	931480	40
21	364064	1010425	939330	71094	931374	39
22	364335	1009681	938473	71208	931268	38
23	364606	1008938	937616	71322	931162	37
24	364877	1008195	936759	71436	931056	36
25	365148	1007453	935903	71550	930950	35
26	365418	1006712	935048	71664	930849	34
27	365689	1005971	934193	71778	930737	33
28	365960	1005231	933339	71893	930631	32
29	366231	1004490	932485	72007	930524	31
30	366501	1003753	931631	72122	930418	30

Min.

Deg. 68

Deg. 21   + | −

m.	Sines.	Logarith.	Differen.	Logarit.	Sines.	
30	366501	1003753	931631	72122	930418	30
31	366772	1003015	930778	72236	930311	29
32	367042	1002277	929926	72351	930204	28
33	367313	1001540	929074	72466	930097	27
34	367583	1000804	928223	72581	929990	26
35	367854	1000068	927373	72696	929884	25
36	368125	999333	926521	72811	929777	24
37	368395	998599	925673	72926	929669	23
38	368665	997865	924824	73041	929562	22
39	368936	997132	923975	73157	929455	21
40	369206	996400	923127	73272	929348	20
41	369476	995668	922280	73388	929240	19
42	369747	994937	921433	73504	929133	18
43	370017	994206	920586	73619	929025	17
44	370287	993476	919741	73735	928917	16
45	370557	991747	918895	73851	928810	15
46	370828	992018	918050	73967	928712	14
47	371098	991290	917206	74084	928594	13
48	371368	990562	916362	74200	928486	12
49	371638	989835	915519	74316	928378	11
50	371908	989109	914676	74433	928270	10
51	372178	988383	913833	94549	928161	9
52	372448	987658	912991	74666	928053	8
53	372718	986933	912150	74783	927945	7
54	372988	986209	911309	74900	927836	6
55	373258	985486	910469	75017	927728	5
56	373527	984763	909629	75134	927619	4
57	373797	984041	908790	75251	927510	3
58	374067	983319	907951	75368	927402	2
59	374337	982599	907113	75486	927293	1
60	374607	981878	906275	75603	927184	0

Min.

Deg. 68

Deg. 22   + | −

m.	Sines.	Logarith.	Differen.	Logarit.	Sines.	
0	374607	981878	906275	75603	927184	60
1	374876	981159	905438	75721	927075	59
2	375146	980440	904601	75838	626966	58
3	375416	979721	903765	75956	926857	57
4	375685	979004	902630	76074	926747	56
5	375955	978286	902094	76192	926638	55
6	376224	977570	901259	76310	916529	54
7	376494	976853	900425	76428	926419	53
8	376763	976138	899591	76547	926310	52
9	377033	975423	898758	76665	926200	51
10	377302	974709	897925	76783	926090	50
11	377571	973995	897093	76902	925980	49
12	377841	973282	896261	77021	925871	48
13	378110	972569	895430	77140	925761	47
14	378379	971857	894599	77259	925651	46
15	378649	971146	893769	77378	925541	45
16	378918	970435	892939	77497	925430	44
17	379187	969735	892120	77616	925320	43
18	379456	969016	891281	77735	925210	42
19	379725	968307	890453	77854	925099	41
20	379994	967599	889625	77974	924989	40
21	380263	966891	888798	78093	924878	39
22	380532	966184	887971	78213	924768	38
23	380801	965477	887145	78332	924657	37
24	381070	964771	886319	78452	924546	36
25	381339	964065	885493	78572	924435	35
26	381608	963360	884668	78692	924324	34
27	381877	962656	883844	78810	924213	33
28	382146	961952	883020	78933	914102	32
29	382415	961249	882196	79053	923991	31
30	382683	960547	881373	79174	923879	30

Min.

Deg. 67

Deg. 22   +|−

m.	Sines.	Logarith.	Differē.	Logarit.	Sines.	
30	382683	960547	881373	79174	923879	30
31	382952	959845	880551	79194	923768	29
32	383221	659143	879728	79415	923657	28
33	383489	958443	878907	79536	923545	27
34	383758	957742	878086	79656	923434	26
35	384027	957043	877265	79777	923322	25
36	384195	956344	876445	79898	923210	24
37	384564	955645	875626	80019	923098	23
38	384832	954947	874806	80141	922987	22
39	385101	954250	873988	80262	922875	21
40	385369	953553	873170	80383	922762	20
41	385638	952857	872352	80505	922650	19
42	385906	952161	871534	80626	922538	18
43	386174	951466	870718	80748	922426	17
44	386443	950771	869901	80870	922313	16
45	386710	950077	869085	80992	922201	15
46	386979	949384	868270	81114	922088	14
47	387247	948691	667455	81236	921976	13
48	387515	947999	866640	81358	921863	12
49	387784	947307	865826	81481	921750	11
50	388052	946616	865013	81603	921638	10
51	388320	945925	864200	81726	921525	9
52	388588	945235	863388	81848	921412	8
53	388856	944546	862575	81971	921299	7
54	389124	943857	861763	82094	921185	6
55	389392	943169	860952	82217	921072	5
56	389660	942481	860141	82340	920959	4
57	389928	941794	859331	82463	920846	3
58	390195	941107	858521	82586	920732	2
59	390463	940421	857712	82709	920618	1
60	390731	939735	846903	82833	920505	0

Min.

Deg. 67

Deg. 23   +|−

m.	Sines.	Logarit.	Differë.	Logarit.	Sines.	
0	390731	939735	856903	82833	920505	60
1	390999	939050	856094	82956	920391	59
2	331267	937366	855286	83080	920277	58
3	391534	937682	854478	83204	920163	57
4	391802	936999	853671	83327	920050	56
5	392070	936316	852865	83451	919936	55
6	392337	935634	852065	83575	919821	54
7	392605	934952	851252	83699	919707	53
8	392872	934271	850447	83824	919593	52
9	393140	933590	849642	63948	919479	51
10	393407	932910	848837	84073	919364	50
11	393675	932230	848033	84197	919250	49
12	393942	931552	847230	84322	919135	48
13	394209	630873	846426	84447	919021	47
14	394477	930195	845624	84572	918906	46
15	394744	929518	844821	84696	918791	45
16	395011	928841	844019	84821	918676	44
17	395278	928165	843218	84947	918561	43
18	395546	927489	842417	85072	918446	42
19	395813	926814	841617	85197	918331	41
20	396080	926139	840817	85322	918216	40
21	396347	925465	840017	85448	918101	39
22	396614	924791	839218	85574	917986	38
23	396881	924118	838419	85699	917870	37
24	397148	923446	837621	85825	917755	36
25	397415	922774	836823	85951	917639	35
26	397682	922103	836026	86077	917523	34
27	397949	921432	835229	86203	917408	33
28	398215	920761	834432	86329	917292	32
29	398482	920092	833636	86456	917176	31
30	398749	919423	832840	86582	917080	30

Min.

Deg. 66

Deg. 23   + | −

m.	Sines.	Logarit.	Differë.	Logarit.	Sines.	
30	378749	919423	832840	86582	917060	30
31	399016	918754	832045	86709	916944	29
32	339283	918086	831250	86836	916828	28
33	399549	917410	830456	86962	916712	27
34	399816	916751	829662	87089	916595	26
35	400082	916084	828868	87216	916497	25
36	400349	915418	828075	87343	916363	24
37	400616	914753	827283	87470	916246	23
38	400882	914088	826490	87597	916130	22
39	401149	913423	825699	87725	916013	21
40	401415	912750	824907	87952	915896	20
41	401681	912096	824116	87979	915780	19
42	401948	911433	823326	88107	915663	18
43	402214	910771	822536	88235	915546	17
44	402480	910109	825746	88363	915429	16
45	402647	909447	820957	88490	915311	15
46	403013	908786	820168	88619	915194	14
47	403279	908126	819379	88747	915077	13
48	403545	907466	818591	88875	914960	12
49	403811	906807	817804	89003	914842	11
50	404078	906148	817016	89132	914725	10
51	404344	905490	816229	89261	914607	9
52	404610	904832	815443	89389	914489	8
53	404876	904175	814657	89518	914372	7
54	405142	903518	813871	89647	914254	6
55	405407	902862	813086	89776	914136	5
56	405673	902207	812301	89905	914018	4
57	405939	901551	811517	90034	913900	3
58	406205	900897	810733	90163	913782	2
59	406471	900243	809950	90293	913664	1
60	406737	899589	809167	90422	913545	0

Min.

Deg. 66

Deg. 24   + | −

m.	Sines.	Logarit.	Differē.	Logarit.	Sines.	
0	406737	899589	809167	90422	913545	60
1	407001	898934	808384	90552	913427	59
2	407268	898183	807602	90681	913309	58
3	407534	897631	806820	90811	913190	57
4	407799	896980	806039	90941	913072	56
5	408065	896329	805258	91071	912953	55
6	408330	895678	804477	91201	912834	54
7	408596	895028	803697	91331	912715	53
8	408861	894378	802917	91461	912596	52
9	409127	893729	802138	91592	912477	51
10	409392	893081	801358	91722	912358	50
11	409658	892433	800580	91852	912239	49
12	409923	891785	799802	91984	912120	48
13	410188	891138	799024	92114	912001	47
14	410454	890492	798247	92245	911881	46
15	410719	889846	797470	92376	911762	45
16	410984	889200	796693	92507	911642	44
17	411249	888555	795927	92639	911523	43
18	411514	887911	795141	92770	911403	42
19	411779	887267	794366	92901	911283	41
20	412045	886623	793591	93033	911164	40
21	412309	885980	792816	93164	911044	39
22	412575	885338	792042	93296	910924	38
23	412839	884696	791268	93428	910804	37
24	413104	884054	790495	93560	910684	36
25	413369	883413	789722	93692	910563	35
26	413634	882773	788949	93824	910443	34
27	113898	882133	788177	93956	910323	33
28	414164	881493	787405	94088	910202	32
29	414428	880854	786634	94221	910082	31
30	414693	880216	785863	94353	909961	30

Min.

Deg. 65

Deg. 24 + | −

m.	Sines.	Logarit.	Differē.	Logarit.	Sines.	
30	414693	880216	785863	94353	909961	30
31	414958	879578	785092	94486	909841	29
32	415223	878940	784322	94618	909720	28
33	415487	878303	783552	94751	909599	27
34	415752	877667	782782	94884	909478	26
35	416016	877031	782013	95017	909357	25
36	416281	876396	781246	95150	909236	24
37	416545	875760	780476	95283	909115	23
38	416810	875125	779708	95417	908994	22
39	417074	874492	778941	95550	908873	21
40	417338	873857	778174	95684	908751	20
41	417603	873224	777407	95818	908630	19
42	417867	872592	776640	95951	908508	18
43	418131	871959	775874	96085	908389	17
44	418395	871328	775108	96119	908265	16
45	418660	870696	774343	96353	908143	15
46	418924	870066	773578	96487	908021	14
47	419188	869435	772814	66622	907899	13
48	519452	868806	772050	96756	907777	12
49	419716	868176	771286	66890	907655	11
50	419980	867547	770523	97025	907533	10
51	420244	866919	769760	97159	907411	9
52	420508	866291	768997	97295	907289	8
53	420772	865664	768230	97429	907166	7
54	421036	865037	767473	97564	907044	6
55	421300	864411	766711	97699	906922	5
56	421563	863781	765950	97834	906799	4
57	421827	863155	765189	97970	906676	3
58	422091	862534	764429	98105	906553	2
59	422355	861910	763669	98241	906431	1
60	422618	861286	762909	98376	906308	0

Min.

Deg. 65

Deg. 25   +|−

m.	Sines.	Logarit.	Differe.	Logarit.	Sines.	
0	422618	861280	762909	98376	906308	60
1	422882	860662	762150	98512	906185	59
2	423145	860039	761391	98648	906062	58
3	423409	859416	760633	98784	905939	57
4	423672	858794	759874	98920	905815	56
5	423936	858172	759117	99056	905692	55
6	424199	857551	758359	99192	905569	54
7	424463	856931	757602	99328	905445	53
8	424726	856310	756846	99465	905322	52
9	424989	855690	756089	99601	905198	51
10	425253	855071	755333	99738	905075	50
11	425516	854452	754578	99875	904951	49
12	425779	853834	753822	100012	904827	48
13	426042	853216	753067	100149	904703	47
14	426306	852598	752313	100286	904579	46
15	426569	851981	751559	100423	904455	45
16	426832	851365	750805	100560	904331	44
17	427095	850749	750052	100697	904207	43
18	427358	850133	749299	100835	904082	42
19	427620	849518	748546	100972	903958	41
20	427884	848903	747794	101110	903834	40
21	428147	848289	747042	101247	903709	39
22	428410	847675	746290	101385	903585	38
23	428672	847062	745539	101523	903460	37
24	428935	846449	744788	101661	903335	36
25	429198	845837	744037	101790	903210	35
26	429461	845225	743287	101938	903086	34
27	429723	844613	742537	101076	902961	33
28	429980	844002	741788	102215	902836	32
29	430248	843392	741039	102353	902710	31
30	430512	842782	740290	102492	901585	30

Min.

Deg. 64

Deg. 25 +|−

m.	Sines.	Logarit.	Differĕ.	Logarit.	Sines.	
30	430511	842782	740290	102492	902585	30
31	430774	842172	739541	102631	902460	29
32	431036	841563	738793	102770	902335	28
33	431299	840954	738046	102909	902209	27
34	431561	840346	737298	103048	902084	26
35	431823	839718	736551	103187	901958	25
36	432086	839131	735804	103326	901833	24
37	432348	838524	735058	103466	901707	23
38	432610	837917	734312	103605	901581	22
39	432873	837311	733566	103745	901455	21
40	433135	836706	732821	103885	901329	20
41	433397	836101	732076	104025	901203	19
42	433659	835496	731332	104164	901077	18
43	433921	834892	730587	104305	900951	17
44	434183	834288	729843	104445	900824	16
45	434445	833685	729100	104585	900698	15
46	434707	833082	728357	104726	900572	14
47	434969	832480	727614	104866	900445	13
48	435231	831878	726871	105006	900319	12
49	435493	831276	726129	105147	900192	11
50	435755	830675	725387	105288	900065	10
51	436017	830075	724646	105429	899939	9
52	436278	829474	723905	105570	899812	8
53	436540	828875	723164	105711	899685	7
54	436802	828275	722423	105852	899558	6
55	437063	827676	721685	105993	899431	5
56	437325	827070	720943	106135	899303	4
57	437587	826488	720204	106276	899176	3
58	437848	825883	719465	106418	899049	2
59	438110	825285	718726	106559	898921	1
60	438371	824689	717987	106701	898795	0

Min.

Deg. 64

Deg. 26   +|−

m.	Sines.	Logarit.	Differen.	Logarit.	Sines.	
0	438371	824689	717987	106701	898794	60
1	438633	824093	717249	106843	898666	59
2	438894	823497	716511	106985	898539	58
3	439155	822902	715774	107128	898411	57
4	439417	822307	715037	107270	898283	56
5	439678	821712	714300	107412	898155	55
6	439939	821118	713564	107555	898028	54
7	440200	820525	712828	107697	897900	53
8	440462	819932	712092	107840	897771	52
9	440723	819339	711357	187982	897643	51
10	440984	818747	710622	108125	897515	50
11	441245	818155	709887	108268	897387	49
12	441506	817564	709152	108411	897258	48
13	441767	816973	708418	108555	897130	47
14	442028	816382	707684	108698	897001	46
15	442289	815792	706951	108841	896873	45
16	442550	815203	706218	108985	896744	44
17	442810	814613	705485	109128	896615	43
18	443071	814025	704753	109272	896486	42
19	443332	813436	704020	109416	896357	41
20	443593	812848	703289	109560	896228	40
21	443853	812261	702557	109704	896099	39
22	444114	811674	701826	109848	895970	38
23	444375	811087	701095	109992	895841	37
24	444635	810501	700365	110136	895712	36
25	444896	809915	699634	110281	895582	35
26	445156	809330	698904	110425	895453	34
27	445417	808745	698175	110570	895323	33
28	445677	808160	697446	110714	895194	32
29	445937	807576	696717	110860	895064	31
30	446198	806993	695988	111005	894934	30

Min.

Deg. 63

Deg. 26 + | −

m.	Sines.	Logarit.	Differë.	Logarit.	Sines.	
30	446198	806993	695988	111005	894934	30
31	446458	806409	695260	111150	894804	29
32	496718	805827	694532	111295	894675	28
33	446977	805244	693804	111440	894545	27
34	447238	804662	593076	111586	894415	26
35	447499	804081	692349	111731	894284	25
36	447759	803500	691623	111877	894154	24
37	448019	802919	690896	112022	894024	23
38	248279	802339	690170	112168	893894	22
39	448539	801759	689445	112314	893763	21
40	448799	801179	688719	112460	893633	20
41	449059	800600	687994	112607	893502	19
42	449319	800022	687269	112753	893372	18
43	449579	799444	686544	112899	893240	17
44	449839	798866	685820	113046	893110	16
45	450098	798289	685096	113192	892975	15
46	450358	797712	684373	113339	892848	14
47	450618	797135	683649	113486	892717	13
48	450878	796559	682926	113633	892586	12
49	451137	795983	682204	113780	892455	11
50	451397	795408	681482	113927	892323	10
51	451656	794833	680760	114074	892192	9
52	451916	794259	680038	114221	892061	8
53	452175	793685	679317	114368	891929	7
54	452435	793111	678595	114516	891798	6
55	452694	792538	677875	114664	891666	5
56	452953	791965	677154	114811	891534	4
57	453213	791393	676434	114959	891402	3
58	453472	790821	675714	115107	891270	2
59	453731	790250	674994	115255	891138	1
60	453990	789679	674275	115403	891006	0

Min.

Deg. 63

Deg. 27   +|−

m.	Sines.	Logarit.	Differen.	Logarit.	Sines.	
0	453990	789679	674275	115403	891006	60
1	454250	789108	673556	115552	890874	59
2	454509	788538	772837	115700	890742	58
3	454768	787968	672119	115849	890610	57
4	454027	787398	671401	115997	890478	56
5	455286	786829	670683	116146	890345	55
6	455545	786260	669966	116295	890213	54
7	455804	785692	669249	116444	890080	53
8	456063	785124	668532	116593	889948	52
9	456322	784557	667815	116742	889815	51
10	456580	783990	667099	116891	889682	50
11	456839	783423	666383	117040	889549	49
12	457098	782857	665668	117190	889416	48
13	457357	782291	664952	117339	889283	47
14	457615	781726	664237	117489	889150	46
15	457874	782161	663522	117639	889017	45
16	458132	780596	662808	117788	888884	44
17	458391	780032	662094	117938	888750	43
18	458650	779468	661380	118089	888617	42
19	458908	778905	660666	118239	888484	41
20	459160	778342	659953	118389	888350	40
21	459415	777780	659240	118539	888217	39
22	459683	777218	658528	118690	888083	38
23	459942	776656	657815	118841	887949	37
24	460200	776094	657103	118991	887815	36
25	460458	775533	656391	119142	887681	35
26	460716	774973	655680	119293	887547	34
27	460974	774412	654968	119444	887413	33
28	461732	773853	654257	119595	887279	32
29	461491	773293	653547	119747	887145	31
30	461749	772734	652836	119898	887011	30

Min.

Deg. 62

Deg. 27   +|−

m.	Sines.	Logarith.	Differen.	Logarit.	Sines.	
30	461749	772734	652836	119898	887011	30
31	462007	772176	652126	120050	886877	29
32	462265	771617	651416	120201	886742	28
33	462522	771060	650707	120353	886608	27
34	462780	770502	649997	120505	886473	26
35	463038	769945	649289	120657	886338	25
36	463296	769389	648580	120809	886203	24
37	463554	768832	647872	120961	886069	23
38	463811	768277	647164	121113	885934	22
39	464069	767721	646456	121265	885799	21
40	464327	767166	645748	121418	885664	20
41	464584	766612	645041	121570	885529	19
42	464842	766057	644334	121723	885394	18
43	465099	765503	643628	121876	885258	17
44	465357	764950	641921	122029	885123	16
45	465614	764397	642215	122182	884988	15
46	465872	763844	641509	122335	884852	14
47	466129	763292	640804	122488	884716	13
48	466387	762740	640099	122641	884581	12
49	466644	762188	639394	122795	884445	11
50	466901	762637	638689	122948	884309	10
51	467158	761087	637985	123102	884174	9
52	467416	760536	637280	123255	884038	8
53	467673	759986	636577	123409	883902	7
54	467930	759437	635873	123563	883766	6
55	468187	758887	635270	123717	883629	5
56	468444	758339	634467	123871	883493	4
57	468701	757790	633764	124026	883357	3
58	468958	757242	633062	124180	883220	2
59	469215	756694	632360	124335	883084	1
60	469472	756147	631658	124489	882948	0

Min.

H   **Deg. 62**

Deg. 28   +|−

m.	Sines.	Logarith.	Differen.	Logarit.	Sines.	
0	469472	756147	631658	124489	881948	60
1	469728	755600	630956	124644	882811	59
2	469985	755054	630255	124799	882674	58
3	470242	754508	629553	124954	882537	57
4	470499	653962	628853	125109	882401	56
5	470755	753416	628152	125264	882264	55
6	471012	752871	627452	125420	882127	54
7	471268	752327	626752	125575	881990	53
8	471525	751783	626052	125730	881853	52
9	471781	751239	625373	125886	881715	51
10	472038	750695	624654	126041	881578	50
11	472294	750152	623955	126197	881441	49
12	472551	749610	623256	126353	881303	48
13	472807	749067	622558	126509	881166	47
14	473063	748525	622860	126661	881028	46
15	473320	747984	621162	126822	880891	45
16	473576	747443	620465	126978	880753	44
17	473832	746902	619768	127134	880615	43
18	474088	746362	619071	127291	880477	42
19	474344	745822	618374	127448	880339	41
20	474600	745282	617677	127604	880201	40
21	474856	744743	616981	127761	880063	39
22	475112	744204	616285	127918	879925	38
23	475368	743665	615590	128075	879787	37
24	475624	743127	614894	128233	879649	36
25	475880	742589	614199	128390	879580	35
26	476136	742052	613506	128547	879372	34
27	476392	741515	612810	128705	879233	33
28	476647	740978	612115	128863	879095	32
29	476907	740442	611421	129020	878956	31
30	477159	739906	610727	129178	878817	30

Min.

Deg. 61

Deg. 28   +|−

m.	Sines.	Logarith.	Differen.	Logarit.	Sines.	
30	477159	739906	610727	129178	878817	30
31	477414	739370	610034	129336	878678	29
32	477670	738835	609340	129494	878539	28
33	477925	738300	608647	129653	878400	27
34	478181	737766	608955	129811	878261	26
35	478436	737232	607262	129969	878122	25
36	478692	736698	606570	130128	877983	24
37	478947	736165	605878	130286	877844	23
38	479203	735632	605186	130445	877704	22
39	479458	735099	604495	130604	877565	21
40	479713	734567	603804	130763	877425	20
41	479968	734035	603113	130922	877286	19
42	480224	733503	602422	131081	877146	18
43	480479	732972	601731	131241	877006	17
44	480734	732441	601041	131400	876867	16
45	480989	731911	600351	131560	876727	15
46	481244	731381	599662	131719	876587	14
47	481499	730851	598972	131879	876447	13
48	481754	730322	598283	132039	876307	12
49	482009	729793	597594	132199	876166	11
50	482263	729264	596906	132359	876026	10
51	482518	728736	596217	132519	875886	9
52	482773	728208	595529	132680	875746	8
53	483028	727681	594841	132840	875605	7
54	483282	727154	594153	133000	875465	6
55	483537	726627	593466	133161	875324	5
56	483792	726101	592779	133322	875183	4
57	484046	725575	592092	133483	875042	3
58	484301	725049	591405	133644	874902	2
59	484555	724524	590715	133805	874161	1
60	484810	723999	590033	133966	874620	0

Min.

H 2   **Deg. 61**

Deg. 29   +|−

m.	Sines.	Logarith.	Differen.	Logarit.	Sines.	
0	484810	723999	590033	133966	875620	60
1	485064	723474	589347	134127	874479	59
2	485318	732950	588661	134289	874338	58
3	485573	722426	587976	134450	874196	57
4	485827	921903	587291	134612	874055	56
5	486081	721379	586606	134774	873914	55
6	486335	720857	585921	134935	873772	54
7	486589	720334	585237	135097	873631	53
8	486844	719812	574553	135259	873489	52
9	487098	719290	583869	135422	873347	51
10	487352	718769	583185	135584	873206	50
11	487606	718248	582501	135746	873064	49
12	487860	717727	581818	135909	872922	48
13	488113	717207	581135	135071	872780	47
14	488367	716687	580453	136234	872638	46
15	488621	716167	579770	136397	872496	45
16	488875	715648	579088	136560	872354	44
17	489129	715129	578406	136723	872212	43
18	489382	714611	577724	136856	872069	42
19	489636	714092	577043	137050	871927	41
20	489890	713575	576361	137213	871784	40
21	490143	713057	575681	137377	871642	39
22	490397	712540	575000	137540	871499	38
23	490650	712023	574319	137704	871356	37
24	490904	711507	573639	137868	871214	36
25	491157	710991	572659	138032	871071	35
26	491410	710475	572279	138196	870928	34
27	491664	709960	571600	138360	870785	33
28	491917	709445	570920	138524	870642	32
29	492170	708930	570241	138689	870499	31
30	492423	708416	569562	138853	870356	30

Min.

Deg. 60

Deg. 29   +|−

m.	Sines.	Logarith.	Differen.	Logarit.	Sines.	
30	492423	708416	569562	138853	870356	30
31	492677	707902	568884	139018	870212	29
32	492930	707388	568206	139183	870069	28
33	493183	706875	567527	139347	869926	27
34	493436	706362	566850	139512	869782	26
35	493689	705849	566172	139677	869639	25
36	493942	705337	565494	139843	869495	24
37	494195	704825	564817	140008	869351	23
38	494448	704314	564140	140173	869207	22
39	494700	703803	563464	140539	869064	21
40	494953	703292	562787	149304	868920	20
41	495206	702781	562111	140670	868776	19
42	495459	702271	561435	140836	868631	18
43	495711	701761	560759	141002	868487	17
44	495965	701252	560084	141168	868343	16
45	496216	700743	559408	141334	868199	15
46	496469	700234	558733	141500	868054	14
47	496721	699726	558059	141667	867910	13
48	496974	699218	557384	141834	867765	12
49	497226	698710	556709	142000	867620	11
50	497479	698202	556035	142167	867476	10
51	497731	697695	555361	142334	867331	9
52	497983	697189	554687	142501	867187	8
53	498236	696682	554014	143668	867042	7
54	498488	696176	553345	142835	866897	6
55	498740	695670	552668	143003	866752	5
56	498992	695165	551995	143170	866607	4
57	499244	694660	551322	143338	866461	3
58	499496	694155	550650	143505	866316	2
59	499748	693651	559978	143673	866171	1
60	500000	693147	549306	143841	866025	0

Min.

H 3   **Deg. 60**

Deg. 30   +|−

m.	Sines.	Logarith.	Differen.	Logarit.	Sines.	
0	500000	693147	549306	143841	866025	60
1	500252	692643	548634	144009	865880	59
2	500504	692140	547963	144177	865734	58
3	500756	691637	547292	144345	865589	57
4	501007	691134	546621	144514	865443	56
5	501259	690632	645950	144682	865297	55
6	501510	690130	645279	144851	865151	54
7	501762	689628	544609	145019	865005	53
8	502014	689127	543939	145188	864859	52
9	502266	688626	543269	145357	864713	51
10	502517	688125	542599	145526	864567	50
11	502769	687625	541930	145695	864421	49
12	503020	687125	541260	145864	864275	48
13	503271	686625	540591	146034	864128	47
14	503523	686126	539923	146203	863982	46
15	503774	685627	539254	146373	863835	45
16	504025	685128	538586	146543	863689	44
17	504276	684630	537918	146712	863542	43
18	504528	684132	537250	146882	863396	42
19	504779	683635	536582	147052	863249	41
20	505030	683137	535915	147223	863102	40
21	505281	682640	535247	147393	862955	39
22	505532	682144	534580	147563	862808	38
23	505783	681648	533914	147734	862661	37
24	506034	681152	533247	147904	862514	36
25	506285	680656	532581	148075	862366	35
26	506535	680161	431915	148246	862219	34
27	506786	679666	531249	148417	862072	33
28	507037	679171	530583	148588	861924	32
29	507288	678677	529918	148759	861777	31
30	507538	678183	529252	148930	861629	30

Min.

Deg. 59

Deg. 30   +|−

m.	Sines.	Logarith.	Differen.	Logarit.	Sines.	
30	507538	678183	529252	148930	861629	30
31	507789	677689	528587	149102	861481	29
32	508040	677196	527922	149273	861334	28
33	508290	676703	527258	149445	861186	27
34	508541	676210	526593	149616	861038	26
35	508791	675717	525929	149788	860890	25
36	509041	675225	525265	149960	860742	24
37	509292	674734	524601	150132	860594	23
38	509542	674242	523938	150305	860446	22
39	509792	673751	523274	150477	860297	21
40	510043	673261	522611	150649	860149	20
41	510293	672770	521948	150822	860001	19
42	510543	672280	521285	150995	859852	18
43	510793	671790	520623	151167	859704	17
44	511043	671301	519961	151340	859555	16
45	511293	670812	519299	151513	859406	15
46	511543	670323	518637	151686	859258	14
47	511793	669835	517975	151860	859109	13
48	412043	669347	517314	152033	858960	12
49	512293	668859	516652	152206	858811	11
50	512542	668371	515991	152380	858662	10
51	512792	667884	515330	152554	858513	9
52	513042	667397	514670	152727	858363	8
53	513292	666911	514009	252901	858214	7
54	513541	666425	513349	153075	858065	6
55	513791	665939	512689	153250	857915	5
56	514010	665453	512029	153424	857766	4
57	514290	664968	511370	153598	857616	3
58	514539	664483	510710	153773	857467	2
59	514789	663998	510051	153947	857317	1
60	515038	663514	509392	154122	857167	0

Min.

H 4   **Deg. 59**

Deg. 31   +|—

m.	Sines.	Logarith.	Differen.	Logarit.	Sines.	
0	515038	663514	509392	154122	857167	60
1	515287	663030	508733	154297	857017	59
2	515537	662546	508074	154472	856868	58
3	515786	662063	507416	154647	856718	57
4	516035	661580	506758	154822	856567	56
5	516284	661097	506100	154998	856417	55
6	516533	660615	505442	155173	856267	54
7	516782	660133	504784	155348	856117	53
8	517031	659651	504127	155524	855966	52
9	517280	659170	503470	155700	855816	51
10	517529	658689	502813	155876	855665	50
11	517778	658208	502156	156052	855515	49
12	518027	657727	501500	156228	855364	48
13	518276	657247	500843	156404	855214	47
14	518525	656768	500187	156580	855063	46
15	518773	656288	499531	156757	854912	45
16	519022	655809	498875	156933	854761	44
17	519271	655330	498220	157110	854610	43
18	519519	654851	497564	157287	854459	42
19	519767	654373	496909	157464	854308	41
20	520016	653895	496254	157641	854156	40
21	520265	653417	495599	157818	854005	39
22	520513	652940	494945	157995	853854	38
23	520761	652463	494290	158172	853702	37
24	521010	651986	493636	158350	853551	36
25	521258	651510	492982	158528	853399	35
26	521506	651034	492328	128705	853248	34
27	521754	650558	491675	158883	853096	33
28	522002	650083	491021	159061	852944	32
29	522251	649607	490368	159239	852792	31
30	522499	649133	489715	159418	852648	30

Min.

Deg. 58

Deg. 31   +|−

m.	Sines.	Logarith.	Differen.	Logarit.	Sines.	
30	522499	649133	489715	159418	852640	30
31	522747	648658	489062	159596	852488	29
32	522995	648184	488410	159774	852336	28
33	523242	647710	487757	159953	852184	27
34	523490	647237	487105	160132	852032	26
35	523738	646763	486753	160310	852879	25
36	523980	646290	485801	160489	851727	24
37	524234	645818	485149	160668	851574	23
38	524481	645345	484498	160847	851422	22
39	524729	644873	483846	161027	851269	21
40	524977	644401	483195	161206	851117	20
41	525224	643030	482544	161385	850964	19
42	525472	643459	481894	161565	850811	18
43	525719	642988	481243	161745	850658	17
44	525966	642517	480593	161925	850505	16
45	526214	642047	479943	162105	850392	15
46	526461	641577	479293	162285	850199	14
47	526708	641108	478643	162465	850046	13
48	526956	640538	477993	162645	849893	12
49	527203	640169	477344	162826	849739	11
50	527450	639705	476695	163006	849586	10
51	527697	639232	476046	163187	849433	9
52	527944	638764	475397	163368	849279	8
53	528291	638296	474748	163548	849225	7
54	528438	637829	474099	163729	848972	6
55	528685	637362	473451	163911	848818	5
56	528932	636895	472803	164092	848664	4
57	529179	636428	472155	164273	848510	3
58	529426	635962	472507	164455	848356	2
59	529672	635496	470860	164636	848202	1
60	529919	635030	470223	164818	848048	0

Min.

H 5   **Deg. 58**

Deg. 32   +|−

m.	Sines.	Logarith.	Differen.	Logarit.	Sines.	
0	529919	635030	470212	164818	848040	60
1	530166	634565	469565	165000	847894	59
2	530412	634100	468919	165182	847730	58
3	530659	633635	468272	165364	847585	57
4	530906	633171	467621	165546	847431	56
5	531152	632706	466978	165728	847276	55
6	531398	632243	466332	165911	847122	54
7	531645	631779	465686	166093	846967	53
8	531891	631316	465040	166276	846813	52
9	532138	630953	464394	166458	846658	51
10	532384	630390	463749	166641	846503	50
11	532630	629928	463103	166824	846348	49
12	532876	629466	462458	167007	846193	48
13	533122	629004	461813	167191	846038	47
14	533368	628542	461168	167374	846883	46
15	533614	628081	460523	167558	845728	45
16	533860	627620	459879	167741	845572	44
17	534106	627160	459235	167925	845417	43
18	534352	626700	458590	168109	845262	42
19	534598	626239	457947	168293	845106	41
20	534844	625780	457303	168477	844951	40
21	535090	625320	456659	168661	844795	39
22	535335	624861	456016	168845	844640	38
23	535581	624402	455373	169030	844484	37
24	535827	623944	454730	169214	844328	36
25	536072	623486	454086	169399	844172	35
26	536318	623028	453444	169584	844016	34
27	536563	622570	452802	169768	843860	33
28	536809	622113	452159	169953	843704	32
29	537054	621656	451517	170139	843548	31
30	537300	621199	450875	170324	842391	30

Min.

Deg. 57

Deg. 32   +|—

m.	Sines.	Logarith.	Differen.	Logarit.	Sines.	
30	537300	621199	450875	170324	843391	30
31	537545	620743	450233	170509	843235	29
32	537790	620286	449592	170695	843079	28
33	538035	619831	448950	170880	842922	27
34	538281	619375	448309	171066	842766	26
35	538526	618920	447668	171252	842609	25
36	538771	618465	447027	171438	842452	24
37	539016	618010	446386	171624	842296	23
38	539261	617556	445745	171810	842139	22
39	539501	617101	445105	171997	841982	21
40	539751	616648	444465	172183	841825	20
41	539985	616194	443824	172370	841666	19
42	540240	615741	443184	172556	841511	18
43	540485	615288	442545	172743	841354	17
44	540720	614835	441905	172930	841196	16
45	540974	614383	441266	173117	841039	15
46	541219	613931	440626	173304	841882	14
47	541464	613479	439988	173492	840724	13
48	541708	613027	439348	173679	840567	12
49	541953	612576	938710	173867	840409	11
50	542197	612125	438071	174054	840251	10
51	542442	611675	437433	174242	840093	9
52	542686	611224	436794	174430	839936	8
53	542930	610774	436156	174618	839778	7
54	543174	610325	435519	174806	839620	6
55	543419	609875	434861	174994	839462	5
56	543663	609426	434243	175183	839304	4
57	543907	608977	433606	175371	839146	3
58	544151	608528	432969	175560	838987	2
59	544305	608080	432332	175748	838829	1
60	544639	607632	431695	175937	838671	0

Min.

Deg. 57

Deg. 33　　+|−

m.	Sines.	Logarith.	Differen.	Logarit.	Sines.	
0	544639	607632	431695	175937	838671	60
1	544883	607184	431058	176126	838512	59
2	545127	606737	430421	176315	838354	58
3	545301	606289	429785	176504	838195	57
4	545614	605842	429149	176694	838036	56
5	545858	605396	428513	176883	837878	55
6	546102	604949	427877	177073	837719	54
7	546346	604503	427241	177262	837560	53
8	546580	604057	426605	177452	837401	52
9	546833	603612	425970	177642	837242	51
10	547076	603167	425335	177832	837083	50
11	547320	602722	424699	178022	836924	49
12	547563	602277	424064	178213	836764	48
13	547807	601833	423430	178403	836605	47
14	548050	601389	422725	178594	836416	46
15	548293	600945	422160	178784	836286	45
16	548536	600301	421526	178975	836127	44
17	548780	600058	420892	179166	835967	43
18	549023	599615	420258	179357	835007	42
19	549266	599172	419624	179548	835648	41
20	549509	598730	418990	179739	835488	40
21	549752	598288	418357	179931	835328	39
22	549995	597846	417724	180122	835168	38
23	550238	597401	417090	180314	835008	37
24	550481	596963	416457	180506	834848	36
25	550724	596522	415624	180698	834688	35
26	550966	596081	415192	180890	834527	34
27	551209	595641	414559	181082	834367	33
28	551452	595201	413927	181274	834207	32
29	551694	594761	413290	181466	834046	31
30	551937	594311	412662	181659	833886	30

Min.

Deg. 56

Deg. 33 +|—

m.	Sines.	Logarit.	Differē.	Logarit.	Sines.	
30	551937	594321	412662	181659	833886	30
31	552179	593883	412031	181851	833725	29
32	552422	593444	411400	182044	833565	28
33	552664	593005	410768	182237	833404	27
34	552907	592566	410137	182430	833243	26
35	553149	592127	409504	182623	833082	25
36	553392	591689	408873	182816	832921	24
37	553634	591252	408242	183009	832760	23
38	553876	590814	407611	183203	832599	22
39	554118	590377	406980	183396	832438	21
40	554360	589940	406350	183590	832277	20
41	554602	589504	405720	183784	832115	19
42	554844	589067	405089	183978	831954	18
43	555086	588631	404459	184172	831793	17
44	555328	588195	403829	184366	831631	16
45	555570	587760	403200	184560	831470	15
46	555812	587325	402570	184755	831308	14
47	556054	586890	401940	184949	831146	13
48	556296	586455	401311	185144	830984	12
49	556537	586021	400682	185339	830823	11
50	506779	585587	400053	185534	830661	10
51	557021	585153	399424	185729	830499	9
52	557262	584719	398795	185924	830337	8
53	557504	584286	398167	186119	830175	7
54	557745	583853	397538	186315	830012	6
55	557987	583430	396910	186510	829850	5
56	558228	582988	396282	186706	829688	4
57	558469	582556	395654	186902	829525	3
58	558711	582123	395026	187098	829363	2
59	558952	582692	394398	187294	829200	1
60	559193	581261	393771	187490	829038	0

Min.

Deg. 56

Deg. 34   +|—

m.	Sines.	Logarit.	Differē.	Logarit.	Sines.	
0	559193	581261	393771	187490	829038	60
1	559434	580829	393829	187686	828879	59
2	559675	580399	580399	187882	828712	58
3	559916	579968	579968	188079	828549	57
4	560157	579538	579538	188275	8283886	56
5	560398	579108	579108	188472	828222	55
6	560639	578678	578678	188669	828060	54
7	560880	578248	578248	188866	827897	53
8	561121	577819	577819	189063	827734	52
9	561361	577390	577390	189260	827571	51
10	561602	576961	576961	189458	827407	50
11	561843	576533	576533	189655	827244	49
12	562083	576105	576105	189853	827081	48
13	562324	575677	575677	190051	826912	47
14	562564	575249	575249	190249	826753	46
15	562805	574622	574622	190447	826590	45
16	563025	574395	574395	190645	826426	44
17	563286	573968	573968	190843	826262	43
18	563526	573541	573541	191041	826098	42
19	563766	573115	573115	191240	825934	41
20	564007	572689	572689	191439	825770	40
21	564247	572263	572263	191637	825606	39
22	564487	571838	571838	191836	825442	38
23	564727	571413	571413	192035	825278	37
24	564967	570988	570988	192234	825114	36
25	565207	570563	570563	192433	824949	35
26	565447	570139	570139	192633	824785	34
27	565687	569714	569714	192832	824620	33
28	565927	569291	569291	193032	824456	32
29	566166	508867	508867	193232	824291	31
30	566406	568444	568444	193431	824126	30

Min.

Deg. 55

Deg. 34   +|−

m.	Sines.	Logarit.	Differĕ.	Logarit.	Sines.	
30	566406	568444	375012	193414	824126	30
31	566646	568020	374389	193631	823961	29
32	566886	567598	373766	193831	823796	28
33	567126	567175	373143	194032	823632	27
34	567365	566753	372521	194232	823467	26
35	567604	566331	371898	194433	823301	25
36	567844	565909	371276	194633	823136	24
37	568083	565487	370653	194834	822971	23
38	568323	565066	370031	195035	822806	22
39	568562	564645	369409	195236	822640	21
40	568801	564224	368787	195437	822475	20
41	569040	563804	368165	195638	822310	19
42	569280	563383	367544	195840	822144	18
43	569519	562963	366922	196041	821978	17
44	569758	562544	366301	196243	821813	16
45	569997	561124	365680	196444	821647	15
46	570236	561705	365059	196462	821481	14
47	570475	561286	364438	196848	821315	13
48	570714	560868	363817	197051	821149	12
49	570952	560449	363196	197253	820983	11
50	571191	560031	362576	197455	820817	10
51	571429	559613	361956	197657	820651	9
52	571669	559196	361335	197860	820485	8
53	571907	558778	360715	198063	820318	7
54	572146	558361	360095	198266	820152	6
55	572384	557944	359476	198469	819985	5
56	572623	557528	358856	198672	819819	4
57	572861	557111	358236	198875	819652	3
58	573100	556695	357617	199078	819485	2
59	573338	556279	356998	199282	819319	1
60	573576	555864	356378	199485	819152	0

Min.

Deg. 55

Deg. 35   +|−

m.	Sines.	Logarit.	Differē.	Logarit.	Sines.	
0	573576	555864	356373	199485	819152	60
1	573815	555449	355759	199589	818935	59
2	574053	555034	355140	199863	818818	58
3	574291	554619	354522	200097	818651	57
4	574529	554204	353903	200301	818484	56
5	574762	553790	353185	200505	818317	55
6	575005	553376	352666	200710	818150	54
7	575243	552962	352048	200914	817982	53
8	575481	552549	351430	201119	817815	52
9	575719	552135	350812	201324	817648	51
10	575957	551722	350194	201528	817480	50
11	576195	551310	349576	201733	817313	49
12	576432	550897	348958	201939	817145	48
13	576670	550485	348341	202144	816977	47
14	576908	550073	347724	202349	816809	46
15	577145	549661	347106	202555	816642	45
16	577383	549250	346489	202760	816474	44
17	577620	548838	345872	202966	816306	43
18	577858	548427	345255	203172	816138	42
19	578095	548017	344639	203378	815969	41
20	578332	547606	344022	203084	815801	40
21	578570	547196	343405	203791	815633	39
22	578807	546786	342789	203997	815465	38
23	579044	546376	342173	204204	815296	37
24	579281	545967	341557	204410	815128	36
25	579518	545558	340941	204617	814959	35
26	579755	545148	340325	204824	814791	34
27	579992	544740	339709	205031	814622	33
28	580229	544332	339093	205238	814453	32
29	580466	543923	338478	205446	814284	31
30	580703	543516	337863	205653	814115	30

Min.

Deg. 54

Deg. 35 +|−

m.	Sines.	Logarit.	Differë.	Logarit.	Sines.	
30	580703	543516	337863	205653	814115	30
31	580940	543108	337247	205861	913946	29
32	981177	542700	336642	206068	813777	28
33	581413	542293	336017	206276	813608	27
34	581650	541886	335402	206484	813439	26
35	581836	541480	334788	206692	813270	25
36	582123	541073	334173	206900	813101	24
37	582359	540667	333559	207108	812931	23
38	582596	540161	332944	207317	812762	22
39	582832	539855	332230	207525	812592	21
40	583069	539450	331716	207734	812423	20
41	583305	539045	331102	207943	812253	19
42	583541	538640	330488	208152	812083	18
43	583777	538235	329874	208361	811914	17
44	584014	537831	329260	208570	811744	16
45	584250	537427	328647	208780	811574	15
46	584486	537023	328033	208989	811404	14
47	584722	536619	327420	209199	811234	13
48	584958	536215	326807	209408	811064	12
49	585194	535812	326193	209618	810894	11
50	585429	535409	325581	209828	810723	10
51	585665	535007	324968	210038	810553	9
52	585901	534604	324356	210249	810483	8
53	586137	534202	323743	210459	810212	7
54	586372	533800	323131	210669	810042	6
55	586608	533398	322518	211880	809871	5
56	586344	532997	321906	211091	809700	4
57	587079	532596	321294	211302	809530	3
58	587314	532195	320682	211513	809359	2
59	587550	531794	320070	211724	809188	1
60	587785	531393	319458	211935	809017	0

Min.

Deg. 54

Deg. 36   +|−

m.	Sines.	Logarit.	Differĕ.	Logarit.	Sines.	
0	587785	531393	319458	211935	808017	60
1	588020	530993	318846	212147	808946	59
2	588256	530593	318235	212358	808675	58
3	588491	530193	317624	212570	808504	57
4	588726	529794	317012	212782	808333	56
5	588961	539395	316401	212994	808161	55
6	589196	538996	315790	213206	807990	54
7	589431	528597	315179	213418	807818	53
8	589666	528198	314568	213630	807647	52
9	589901	527800	313957	213843	807475	51
10	590136	527402	313347	214055	807304	50
11	590371	527004	312736	214268	807132	49
12	590606	526607	312126	214481	806960	48
13	590840	526209	311515	214694	806788	47
14	591075	525812	310905	214907	806617	46
15	591310	525415	310295	215120	806445	45
16	591544	525019	309685	215333	806373	44
17	591779	524622	309075	215547	806100	43
18	592013	524226	308466	215760	805928	42
19	592248	523830	307856	215974	805756	41
20	592482	523434	307241	216188	805584	40
21	592716	523039	306637	216402	805412	39
22	892950	522644	306028	216611	805239	38
23	593185	522249	305419	216830	805066	37
24	593419	521854	304810	217045	804894	36
25	593653	521460	304201	217259	804721	35
26	593887	521066	303592	217474	804548	34
27	594121	520672	302983	217689	804376	33
28	594555	520278	302375	217904	804203	32
29	594589	519885	301766	218119	804030	31
30	594823	519491	301158	218334	803857	30

Min.

Deg. 53

Deg. 36 +|−

m.	Sines.	Logarit.	Differē.	Logarit.	Sines.	
30	594823	519492	301158	218334	803857	30
31	595057	519099	300549	218549	803684	29
32	595290	518706	299941	218765	803511	28
33	595524	518313	299333	218980	803337	27
34	595758	517921	298725	219196	803164	26
35	595991	517529	298117	219412	802991	25
36	596225	517137	297509	219628	802817	24
37	596458	516746	296902	219844	802644	23
38	596692	516354	296294	220060	802470	22
39	596925	515963	295687	220276	802297	21
40	597150	515572	295079	220493	802123	20
41	597392	515182	294472	220710	801949	19
42	597625	514791	293865	220926	801776	18
43	597858	514401	293258	221143	801602	17
44	598091	514011	292651	221360	801428	16
45	598325	313622	292044	221577	801254	15
46	598558	513232	291437	221795	801080	14
47	598791	512843	290831	222012	800906	13
48	599024	512454	290224	222230	800731	12
49	599257	512065	289618	222447	800557	11
50	599489	511677	289012	222665	800383	10
51	599722	511289	288406	222883	800208	9
52	599955	510901	287799	223101	800034	8
53	600188	510513	289193	223319	799859	7
54	600420	510125	286588	223538	799685	6
55	600653	509738	285982	223756	799510	5
56	600885	509351	285376	223975	799335	4
57	601118	508964	284771	224193	799160	3
58	601350	508577	284165	224412	798985	2
59	601583	508191	284560	224631	798810	1
60	601815	507805	282954	224851	798635	0

Min.

Deg. 53

Deg. 37   +|—

m.	Sines.	Logarit.	Differen.	Logarit.	Sines.	
0	601815	507805	282954	224851	798635	60
1	602047	507419	282349	225070	798465	59
2	602280	507033	281744	225289	798285	58
3	602512	506648	281139	225509	798110	57
4	602744	506263	280534	225728	797935	56
5	602976	505878	279929	225948	797759	55
6	603208	505493	279325	226168	797584	54
7	603440	505108	278720	226388	797408	53
8	603672	504724	278116	226608	797233	52
9	603904	504340	277521	226829	797057	51
10	604136	503956	276907	227049	796881	50
11	604367	503573	276303	227270	796706	49
12	604599	503189	275699	227490	796530	48
13	604831	502806	275095	227711	796354	47
14	605062	502423	274491	227932	796178	46
15	605294	502041	273887	228153	796002	45
16	605525	501658	273284	228375	795826	44
17	605757	501276	272680	228596	795650	43
18	605988	500894	272076	228818	795473	42
19	606220	500512	271473	229039	795297	41
20	606451	500131	270870	229261	795121	40
21	606682	499750	270266	229483	794944	39
22	606914	499369	269663	229705	794768	38
23	607145	498988	269060	229928	794591	37
24	607376	498607	268457	230150	794415	36
25	607607	498227	267854	230372	794238	35
26	607838	497847	267252	230595	794061	34
27	608069	497467	266649	230818	793884	33
28	608300	497087	266047	231041	793707	32
29	608531	496708	265444	231263	793530	31
30	608761	496329	264842	231487	793353	30

Min.

Deg. 52

Deg. 37   +|−

m.	Sines.	Logarit.	Differen.	Logarit.	Sines.	
30	608761	496329	264842	231487	793353	30
31	608992	495950	264240	231710	793176	29
32	609223	495571	263638	231933	792999	28
33	609454	495192	263036	232157	792822	27
34	609684	494814	262434	232380	792644	26
35	609915	494436	261833	232604	792467	25
36	610145	494058	261230	232828	792290	24
37	610376	493681	260628	233052	792112	23
38	610606	493303	260027	233276	791934	22
39	610836	492926	259425	233501	791757	21
40	611067	492549	258824	233729	791579	20
41	611297	492172	258222	233950	971401	19
42	611527	491796	257621	234175	791223	18
43	611757	491419	257020	234400	791046	17
44	611987	491043	256419	234625	790868	16
45	612217	490668	255818	233850	790690	15
46	612447	490292	255217	235075	790511	14
47	612677	489917	254616	235300	790333	13
48	612907	489542	254016	235526	790155	12
49	613137	489167	258415	235752	789977	11
50	613367	488792	252814	235978	789798	10
51	613596	488418	252214	236204	789620	9
52	613826	488043	251614	236430	789441	8
53	614056	487669	251014	236656	789263	7
54	614285	487296	250413	236882	789084	6
55	614515	486922	249813	237109	788905	5
56	614744	486549	249213	237335	788727	4
57	614975	486176	248614	237562	788548	3
58	615203	485803	248014	237789	788369	2
59	615432	485430	247414	238016	788190	1
60	615662	485058	246814	238243	788011	0

Min.

Deg. 52

Deg. 38   + | −

m.	Sines.	Logarit.	Differē.	Logarit.	Sines.	
0	615661	485058	246814	238243	788011	60
1	615891	484686	246215	238471	787832	59
2	616120	484314	245615	238699	787652	58
3	616349	483942	245016	238926	787473	57
4	616578	583570	244417	239153	787294	56
5	616807	483199	243817	239381	787114	55
6	617026	482828	243218	239609	786935	54
7	617265	482457	242619	239838	786755	53
8	617494	482086	242020	240066	786576	52
9	617722	481716	241421	240294	786396	51
10	617951	481346	240823	240523	786217	50
11	618180	480976	240224	240751	786037	49
12	618408	480606	239625	240980	785857	48
13	618637	480236	239027	241209	785677	47
14	618866	479867	238428	241438	785497	46
15	619094	479498	237830	241668	785317	45
16	619322	479129	237232	241897	785137	44
17	619551	478760	236633	242127	784952	43
18	619779	478392	236035	242356	784776	42
19	620007	478024	235437	242586	784596	41
20	620236	477656	234840	242816	784416	40
21	620464	477288	234242	243046	784235	39
22	620692	476920	233644	243276	784055	38
23	620920	476553	233046	243507	783874	37
24	621148	476186	232449	243737	783693	36
25	621376	475819	231851	243968	783513	35
26	621604	474452	231254	244199	783332	34
27	621831	475086	230656	244429	783151	33
28	622059	474720	230059	244660	782070	32
29	622287	474354	229462	244892	782789	31
30	622515	473988	228865	245123	782608	30

Min.

Deg. 51

Deg. 38 +|−

m.	Sines.	Logarit.	Differen.	Logarit.	Sines.	
30	622515	473988	228865	245123	782608	30
31	622742	473622	228268	245354	782427	29
32	622970	473257	227671	245586	782246	28
33	623197	472892	227074	245818	782065	27
34	623425	472527	226477	246050	781883	26
35	623652	472164	225880	246282	781702	25
36	623880	471798	225284	246514	781520	24
37	624107	471434	234688	246746	781339	23
38	624334	471069	224091	246978	781157	22
39	624561	470706	223495	247211	780976	21
40	624788	470341	222898	247444	780764	20
41	625016	469978	222302	247676	780912	19
42	625243	469615	221706	247909	780430	18
43	625469	469252	221110	248143	780248	17
44	625697	469889	220514	248376	780066	16
45	625923	468527	219918	248609	779884	15
46	626150	468164	219322	248843	779702	14
47	626377	467802	218726	249076	779520	13
48	626604	467440	218130	249310	779338	12
49	626830	467979	217534	249544	779156	11
50	627057	466717	216939	249778	778973	10
51	627284	466356	216343	250013	778761	9
52	627510	465995	215748	250247	778608	8
53	627737	465634	215153	250481	778426	7
54	627963	465274	214557	250716	778243	6
55	628189	464913	213962	250951	778060	5
56	628416	464553	213367	251186	778878	4
57	628642	464193	212772	251421	777695	3
58	628868	463833	212177	251656	777512	2
59	629094	463974	211582	251891	777329	1
60	629320	463115	210988	252127	776146	0

Min.

Deg. 51

Deg. 39   +|−

m.	Sines.	Logarit.	Differen.	Logarit.	Sines.	
0	629320	463115	210988	252127	777146	60
1	629546	462755	210393	252363	776963	59
2	629772	462397	1209798	252598	776780	58
3	629998	462038	209204	252834	776596	57
4	630224	461679	208609	253070	776413	56
5	630450	461321	208015	253306	776230	55
6	630676	460963	207420	253543	776046	54
7	630902	460605	206826	253779	775863	53
8	631127	460248	206232	254016	775679	52
9	631353	459890	205638	254253	775496	51
10	631578	459533	205043	254489	775312	50
11	631804	459176	204449	254726	775128	49
12	632029	458819	203855	254964	774944	48
13	632155	458463	203262	255201	774761	47
14	632480	458106	202668	255438	774577	46
15	632705	457750	202074	255676	774393	45
16	632930	457394	201481	255914	774208	44
17	633156	457039	200887	256152	774024	43
18	633381	456683	200293	256390	773840	42
19	633606	456328	199700	256628	773565	41
20	633831	455973	199107	256866	773472	40
21	634056	455628	198513	257105	773287	39
22	634281	455263	197920	257343	773103	38
23	634506	454909	197327	257582	772918	37
24	634731	454555	196734	257821	772734	36
25	634955	454201	196141	258060	772549	35
26	635180	453847	195548	258299	772364	34
27	635405	453493	194955	258538	772179	33
28	635629	453140	194362	258778	771994	32
29	635854	452787	193769	259017	771810	31
30	636078	452434	193177	259257	771625	30

Min.

Deg. 50

Deg. 39 + | −

m.	Sines.	Logarit.	Differē.	Logarit.	Sines.	
30	636078	452434	193177	259257	771625	30
31	636303	452081	192584	259497	771439	29
32	636527	451728	191991	259737	771254	28
33	636751	451376	191399	259977	771069	27
34	636976	451024	190806	260217	770884	26
35	637200	450672	190214	260458	770699	25
36	637424	450320	189622	260698	770513	24
37	637648	449968	189029	260939	770328	23
38	637872	449617	188437	261180	770142	22
39	638096	449266	187845	261421	769957	21
40	638320	448915	187253	261662	769771	20
41	638544	448564	186661	261903	769585	19
42	638768	448214	186069	262145	769399	18
43	638992	447864	185477	262386	769214	17
44	639215	447514	184885	262628	769028	16
45	639439	447164	184294	262870	768842	15
46	639663	446814	183702	263112	768655	14
47	639886	446465	183110	263354	768469	13
48	640120	446115	182519	263596	768283	12
49	640335	445766	181928	263838	768097	11
50	640557	445418	181336	264081	767911	10
51	640980	445069	180745	264324	767725	9
52	641003	444721	180154	264567	767538	8
53	641226	444372	179562	264810	767352	7
54	641450	444024	178971	265053	767165	6
55	641673	443677	178380	265296	766979	5
56	641896	443329	177789	265540	766792	4
57	642119	442982	177198	265783	766605	3
58	642342	442634	176607	266027	766418	2
59	642565	442287	176017	266271	766231	1
60	642778	441941	175426	266515	766044	0

Min.

I **Deg. 50**

Deg. 40   +|—

m.	Sines.	Logarith.	Differen.	Logarit.	Sines.	
0	642788	441941	175426	175426	766044	60
1	643010	441594	174835	174835	765857	59
2	643233	441248	174245	174245	765670	58
3	643456	440902	173654	173654	765483	57
4	643678	440556	173064	173064	765296	56
5	643901	440210	172473	172473	765109	55
6	644124	439865	171883	171883	764921	54
7	644346	439519	171293	171293	764734	53
8	644568	439174	170702	170702	764547	52
9	644791	438829	170112	170112	764359	51
10	645013	438484	169522	169522	764171	50
11	645235	438140	168931	168931	763984	49
12	645458	437795	168341	168341	763796	48
13	645680	437451	167751	167751	763608	47
14	645902	437107	167161	167161	763420	46
15	646124	436764	166571	166571	763232	45
16	646346	436420	165982	165982	763044	44
17	646568	436077	165392	165392	762856	43
18	646790	435734	164802	164802	762668	42
19	647012	435391	164212	164212	762480	41
20	647233	435048	163623	163623	762292	40
21	647455	434706	163033	163033	762104	39
22	647677	434363	162444	162444	791915	38
23	647898	434021	161854	161854	761727	37
24	648120	433679	161265	161265	761538	36
25	648341	433338	160675	160675	761350	35
26	648563	432996	160086	160086	761161	34
27	648784	432655	159497	159497	760972	33
28	649005	432314	158908	158908	760784	32
29	649227	431973	158319	158319	760595	31
30	649448	431632	157730	157730	760406	30

Min.

Deg. 49

Deg. 40 $+|-$

m.	Sines.	Logarith.	Differen.	Logarit.	Sines.	
30	649448	431632	157730	273903	760406	30
31	649669	431292	157141	274151	760217	29
32	649890	430952	156552	274400	760028	28
33	650111	430612	155963	274649	759839	27
34	650332	430272	155374	274898	759650	26
35	650553	429932	154785	275147	759461	25
36	650774	429592	154196	275396	759271	24
37	650995	429253	153608	275645	759082	23
38	651216	428914	153019	275895	758892	22
39	651436	428575	152430	276145	758703	21
40	651657	428236	151842	276394	758514	20
41	651878	427898	151253	276644	758324	19
42	652098	427560	150665	278895	758134	18
43	652319	427222	150077	277145	757945	17
44	652539	426884	149488	277395	757755	16
45	652760	426546	148900	277646	757565	15
46	652980	426208	148312	277897	757375	14
47	653200	425871	147724	278147	757185	13
48	653421	425534	147136	278398	756995	12
49	653641	425197	146548	278650	756805	11
50	653861	424860	145960	278901	756615	10
51	654081	424524	145372	279152	756425	9
52	654301	424188	144784	279404	756234	8
53	654521	423852	144196	279656	756044	7
54	654741	423516	143608	279907	755853	6
55	654961	423180	143020	280159	755663	5
56	655180	422844	142433	280412	755472	4
57	655400	422509	141845	280664	755282	3
58	655620	422174	141257	280917	755091	2
59	655839	421839	140670	281169	754900	1
60	656059	421504	140082	281422	754710	0

Min.

I 2 # Deg. 49

Deg. 41   +|−

m.	Sines.	Logarit.	Differē.	Logarit.	Sines.	
0	656059	421504	140082	281422	754710	60
1	656278	421170	139495	281675	754519	59
2	656498	420835	138907	281928	754328	58
3	656717	420501	138320	282181	754137	57
4	656937	420167	137732	282435	753946	56
5	657156	419833	137145	282688	753755	55
6	657375	419500	136558	282942	753563	54
7	657594	419167	135971	283196	753372	53
8	657814	418833	135384	283450	753181	52
9	658033	418501	134797	283704	752989	51
10	658252	418168	134210	283958	752798	50
11	658470	417835	133623	284213	752606	49
12	658689	417503	133036	284467	752415	48
13	658908	417171	132449	284722	752223	47
14	659127	416839	131862	284977	752032	46
15	659346	416507	131275	285232	751840	45
16	659564	416175	130688	285487	751684	44
17	659783	415844	130102	285742	751456	43
18	660002	415513	129515	285998	751264	42
19	660220	415182	128928	286253	751072	41
20	660439	414851	128342	286509	750880	40
21	660657	414520	127755	286765	750688	39
22	660875	414190	127169	287021	750496	38
23	661094	413860	126582	287277	750303	37
24	661312	413530	125996	287534	750111	36
25	661530	413200	125410	287790	749919	35
26	661748	412870	124823	288047	749726	34
27	661966	412541	124237	288304	749534	33
28	662184	412211	123651	288561	749341	32
29	662402	411882	123064	288818	749148	31
30	662620	411553	122478	289075	748956	30

Min.

Deg. 48

Deg. 41   + | −

m.	Sines.	Logarit.	Differē.	Logarit.	Sines.	
30	662620	411553	122478	289075	748956	30
31	662838	411225	121892	289333	748763	29
32	663056	410896	121306	289590	748570	28
33	663273	410568	120720	289848	748377	27
34	663491	410240	120134	290106	748184	26
35	663709	409912	119548	290364	747991	25
36	663926	409584	118962	290622	747798	24
37	664144	409257	118376	290880	747605	23
38	664361	408929	117790	291139	747412	22
39	664579	408602	117284	291398	747218	21
40	664796	408275	116619	291656	747025	20
41	665013	407948	116033	291915	746832	19
42	665230	407622	115447	292174	746638	18
43	665448	407295	114861	292434	746445	17
44	665665	406969	114276	292693	746251	16
45	665882	406643	113690	292953	746057	15
46	666098	406317	113105	293212	745864	14
47	666316	405992	112519	293472	745670	13
48	966532	405666	111934	293732	745476	12
49	666749	405341	111349	293992	745202	11
50	666966	405016	110763	294253	745088	10
51	667183	404691	110178	294513	744894	9
52	667399	404366	109593	294774	744700	8
53	667616	404042	109007	295034	744506	7
54	667833	403717	108422	295295	744312	6
55	668049	403393	107837	295556	744117	5
56	668265	403069	187252	295818	743923	4
57	668482	402746	106667	296079	743728	3
58	668698	402422	106082	296341	743534	2
59	668914	402099	105497	296602	743339	1
60	669131	401776	104912	296864	743145	0

Min.

I 3   Deg. 48

Deg. 42   +|−

m.	Sines.	Logarith.	Differen.	Logarit.	Sines.	
0	669131	401776	104912	296864	743145	60
1	669374	401453	104327	297126	742950	59
2	669563	401130	103742	297388	742755	58
3	669779	400807	103157	297651	742560	57
4	669995	400485	102572	297913	742366	56
5	670211	400163	101987	298176	742171	55
6	670429	399841	101402	298438	741976	54
7	670642	399519	100818	298701	741781	53
8	670858	399197	100233	298964	741586	52
9	671074	398876	99648	199228	741390	51
10	671289	398555	99064	299491	741195	50
11	671505	398233	98479	299754	741000	49
12	671721	397913	97894	300018	740805	48
13	671936	397592	97310	300282	740609	47
14	672151	397271	96725	300546	740414	46
15	672367	396951	96141	300810	740238	45
16	672852	396631	95556	301074	740022	44
17	672797	396311	94972	301339	739827	43
18	673012	395991	94388	301604	739631	42
19	673328	395672	93803	301868	739435	41
20	673443	395352	93219	302133	739239	40
21	673658	395033	92635	302398	739043	39
22	673873	394714	92050	302664	738847	38
23	674087	394395	91466	302929	738651	37
24	674302	394076	90882	303194	738455	36
25	674517	393758	90298	303460	738259	35
26	674732	393440	89713	303726	738063	34
27	674946	393121	89129	303992	727867	33
28	675161	392804	88545	304258	737670	32
29	675376	392486	87961	304525	737474	31
30	675590	392168	87377	504791	737277	30

Min.

Deg. 47

Deg. 42 + | —

m.	Sines.	Logarith.	Differen.	Logarit.	Sines.	
30	675590	392168	87377	304791	737277	30
31	675805	391851	86793	305058	737081	29
32	676019	391534	86209	305324	736884	28
33	676233	391217	85625	305591	736687	27
34	676448	390900	85042	305858	736491	26
35	676662	390583	84458	306126	736294	25
36	676876	390267	83874	306393	736097	24
37	677090	389951	83290	306661	735900	23
38	677304	389635	82706	306928	735703	22
39	677518	389319	82122	907196	735506	21
40	677732	389003	81539	907464	735309	20
41	677946	388688	80955	907733	735112	19
42	678160	388372	80371	308001	734914	18
43	678373	388057	79788	308269	734717	17
44	678597	387742	79204	308538	734520	16
45	678801	387427	78620	308807	734322	15
46	679014	387113	78037	309976	734125	14
47	679228	386798	77453	309345	733927	13
48	679441	386484	76870	309614	733730	12
49	679654	386170	76286	309884	733532	11
50	679898	385856	75703	310153	733334	10
51	680081	385543	75120	310423	733137	9
52	680295	385229	74536	310693	732939	8
53	680508	384916	73953	310963	732741	7
54	680721	384603	73370	311233	732543	6
55	680934	384290	72786	311503	732345	5
56	681147	383977	72203	311774	732147	4
57	681360	383664	71620	312045	731949	3
58	681573	383352	71036	312316	731750	2
59	681786	383040	70453	312587	731552	1
60	681998	382728	69870	312858	731354	0

Min.

I 4 **Deg. 47**

Deg. 43   + | −

m.	Sines.	Logarit.	Differě.	Logarit.	Sines.	
0	681998	382728	69870	312858	731354	60
1	682211	382416	69287	313129	731155	59
2	682424	382104	68703	313401	730957	58
3	682636	381793	68120	313673	730758	57
4	982849	381482	67537	313944	730560	56
5	683061	381170	66954	314216	730361	55
6	683274	380860	66371	314488	730162	54
7	683486	380549	65788	314761	729963	53
8	683698	380238	65205	315033	729765	52
9	683911	379928	64622	315306	729565	51
10	684123	379618	64039	315578	729367	50
11	684335	379307	63456	315851	729168	49
12	684547	378998	62873	316124	758969	48
13	684759	378688	62290	316398	728769	47
14	684971	378378	61707	316671	628570	46
15	685183	378069	61125	316944	728371	45
16	685395	377760	60542	317218	728172	44
17	685607	377451	59959	317492	727972	43
18	685818	377142	59376	317766	727773	42
19	686030	376834	58793	318040	727573	41
20	686242	376525	58210	318315	727374	40
21	686453	376217	57628	318589	727174	39
22	686665	375909	57045	318864	726974	38
23	686876	375601	56462	319139	726775	37
24	687088	375293	55879	319414	726575	36
25	687299	374986	55297	319689	726375	35
26	687510	374679	54714	319964	726175	34
27	687721	374371	54131	320240	725975	33
28	687932	374064	53549	320515	725775	32
29	688144	373758	52966	320791	725575	31
30	688355	373451	52384	321067	725874	30

Min.

Deg. 46

Deg. 43   +|−

m.	Sines.	Logarit.	Differë.	Logarit.	Sines.	
30	688355	373451	52384	321067	725374	30
31	688566	373145	51801	321343	725174	29
32	688776	372838	51219	321620	724974	28
33	688987	372532	50636	321896	724773	27
34	689198	372226	50054	322173	724573	26
35	689409	371921	49471	322449	724372	25
36	689620	371615	48889	322726	724172	24
37	689830	371310	48306	323003	723971	23
38	690041	371004	47724	323281	723770	22
39	690251	370700	47141	323558	723570	21
40	690462	370395	46559	323836	723369	20
41	690672	370090	45976	324114	723168	19
42	690882	369785	45394	324392	722967	18
43	691093	369481	44811	324670	722766	17
44	691303	369177	44229	324948	722565	16
45	691513	368873	43647	325226	722364	15
46	691723	368569	43064	325505	722163	14
47	691933	368266	42482	325783	721961	13
48	692143	367962	41900	326062	721760	12
49	692353	367659	41318	326341	721559	11
50	692563	367356	40735	326620	721357	10
51	692773	367053	40153	326900	721156	9
52	692982	366750	39571	327179	720954	8
53	693192	366448	38989	327459	720753	7
54	693402	366145	38407	327739	720551	6
55	663611	365843	37824	328019	720349	5
56	693821	365541	37242	328299	720148	4
57	694030	365239	36660	328579	719946	3
58	694240	364938	36078	328860	719744	2
59	694449	364636	35496	329140	719542	1
60	694658	364335	34914	329421	719340	0

Min.

Deg. 46

Deg. 44 +|−

m.	Sines.	Logarit.	Differē.	Logarit.	Sines.	
0	694658	364335	34914	329421	719340	60
1	694868	364034	34331	329702	719138	59
2	695077	363733	33749	329983	718935	58
3	695286	363432	33167	330265	718733	57
4	695495	363131	32585	330546	718531	56
5	695704	362831	32003	330828	718329	55
6	695913	362531	31421	331110	718126	54
7	696122	362231	30835	331392	717924	53
8	696330	361931	30257	331674	717721	52
9	696539	361631	29675	331956	717519	51
10	696748	361331	29093	332238	717316	50
11	696959	361032	28511	332521	717113	49
12	697165	360733	27929	332804	716911	48
13	697374	360434	27347	333087	716708	47
14	697582	360135	26765	333370	716505	46
15	697790	359836	26183	333653	716302	45
16	697999	359538	25601	333937	716099	44
17	698207	359239	25019	334220	715896	43
18	698415	358941	24437	334504	715693	42
19	698623	358643	23855	334788	715489	41
20	698832	358345	23273	335072	715286	40
21	699040	358048	22691	335357	715083	39
22	699348	357750	22109	335641	714880	38
23	699455	357453	21527	335926	714676	37
24	699663	357156	20945	336210	714473	36
25	699871	356859	20363	336495	714269	35
26	700079	356562	19782	336781	714065	34
27	700287	356266	19200	337066	713862	33
28	700494	355969	18618	337351	713658	32
29	700702	355673	18036	337637	713454	31
30	700909	355377	17454	337923	713250	30

Min.

Deg. 45

Deg. 44 + | −

m.	Sines.	Logarit.	Differe.	Logarit.	Sines.	
30	700909	355377	17454	337923	713250	30
31	701117	355081	16872	338208	713046	29
32	701324	354785	16290	338495	712842	28
33	701531	354489	15709	338781	712638	27
34	701739	354194	15127	339067	712434	26
35	701946	353899	14545	339354	712230	25
36	702153	353604	13963	339641	712026	24
37	702360	353309	13381	339927	711822	23
38	702567	353014	12800	440225	711617	22
39	702774	352720	12218	340502	711413	21
40	702981	352425	11636	340789	711209	20
41	703188	352131	11054	341077	711004	19
42	703395	351837	10472	341365	710799	18
43	703601	351543	9890	341653	710595	17
44	703808	351249	9309	341941	710390	16
45	704015	350956	8727	342229	710185	15
46	704221	350662	8145	342518	709981	14
47	704428	350369	7563	342806	709776	13
48	704634	350076	6982	343094	709571	12
49	704841	349783	6401	343383	709366	11
50	705047	349491	5818	343673	709161	10
51	705253	349198	5236	343962	708956	9
52	705459	348906	4654	344252	708750	8
53	705665	348614	4073	344541	708545	7
54	705872	348322	3491	344831	708340	6
55	706078	348030	2909	345121	708134	5
56	706284	347738	2327	345411	707929	4
57	706489	347447	1745	345701	707724	3
58	706695	347156	1164	345992	707518	2
59	706901	346864	582	346283	707312	1
60	707107	346573	0	346573	707107	0

Min.

Deg. 45

Place this figure next after the tables.

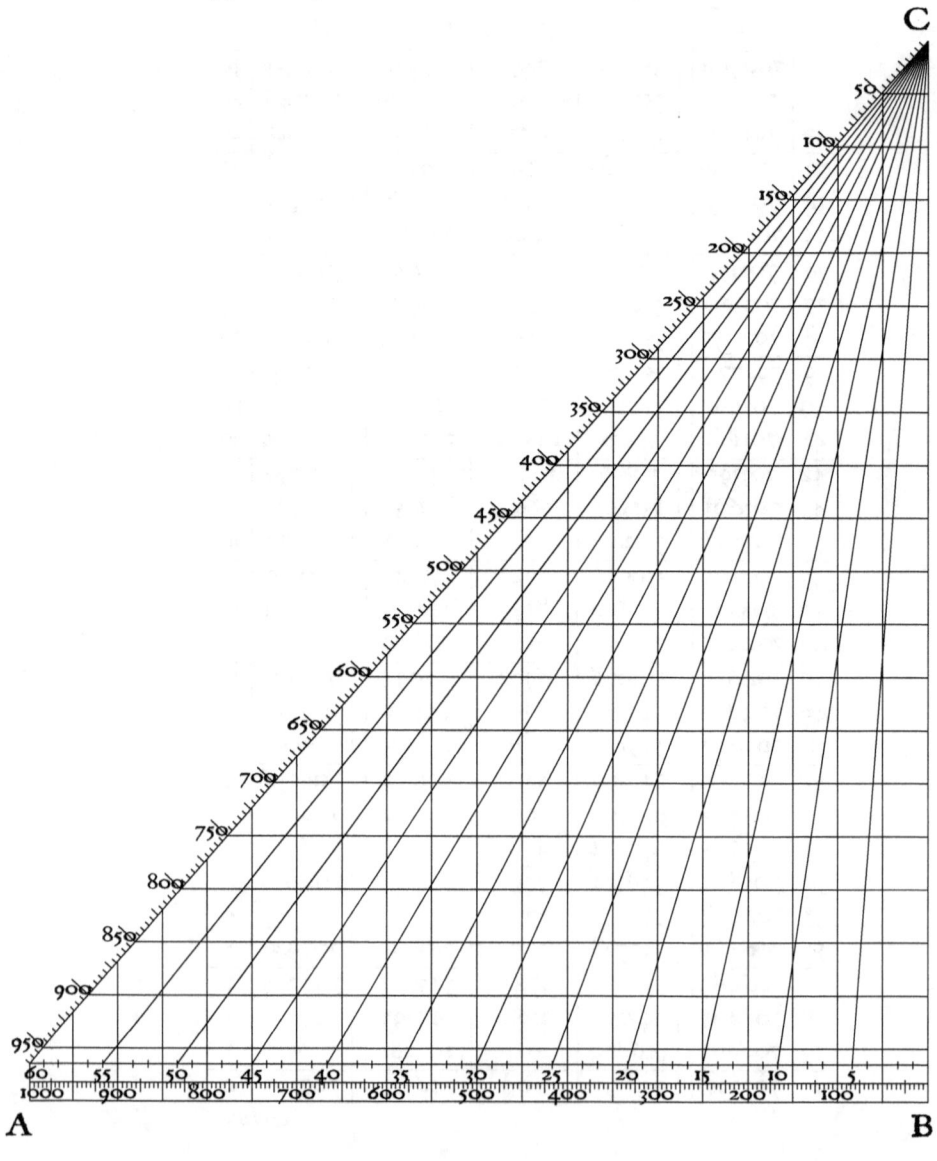

(1)

THE VSE OF THE TRI-
angular Table for the finding of
the part Proportionall, penned
by Henry Brigges.

HE compendioufneffe of thefe Tables of *Logarithmes,* cannot be without fome defeft, which is to be fupplied (as in all other Tables) by the part proportionall : that whereas fomtimes the number we defire, is not to be found in the Table, we may by the difference of that number, the number in the Table neereft vnto it, and the proportionall part anfwerable to that difference, haue our defire, fo neere as may be, or is needfull. This proportionall part is had three wayes.

1. Either by the Rule of Proportion, commonly called the Golden Rule : Or
2. By the helpe of this Table of Logarithmes. Or
3. By this Triangular inftrumentall Table.

Of which three wayes the firft is moft exaft, and the laft moft eafie, but not fo exaft as the other two. All thefe wayes hauing three numbers giuen, do helpe vs to finde the fourth proportionall number. The manner of the operaration is beft fhewed by example.

Pag. 81. lin. 15. we haue found that 141766 is the Logarithme of halfe the angle PZS. which number I feeke in the Table, and finde it not; but the two *Logarithmes* next vnto it,

are

(2)

are 141834 and 141667, which anſwere to 12
& 13 minuts aboue 60 degrees: ſo that it is apparant, that the arke which we ſeeke is 60 : 12
and more. Now to finde how much more is to
be added to this number, I take the difference
of the tabular Logarithmes, 167 being firſt of
the three numbers, which before I ſaid muſt
be giuen, and the difference of the two anſwe-
rable tabular arkes 60 ſeconds, which is the ſecond giuen number, and theſe two may fitly
be termed *Tabular differences*. The third giuen
number which we call the *Occurring difference*,
is the difference of the former of the two Tabular Logarithmes, and of this Logarithme
141766 which we ſeeke for: which difference
is 68. Theſe three giuen numbers 167 60. 68
do helpe vs to the part proportionall, to be added to 60 : 12, by euery one of the three former wayes.

1. By the golden rule I multiply the third 68,
by the ſecond 60, and the product 4080 being
diuided by the firſt 167, giueth in the quotient 24 $\frac{1}{2}$ almoſt, ſo that the arke anſwering
to halfe the angle PZS, is found to be 60 : 12,
24$\frac{1}{2}$.

2. By this Table of Logarithmes thus, I take
the Logarithmes of theſe three giuen numbers, ſo neere as the bare Table will affoord,
without any further curious ſearch (becauſe
in this caſe I need not ſeeke to be exact; and
for the ſame cauſe I cut off the two laſt figures
in euery Logarithme) the Logarithmes I find
to be 17928. 5109. 3943. But becauſe the numbers in the Table, to which the Logarithmes
are adioyned, are *Sines*, and I haue found theſe
Logarithmes anſwerably, as if the giuen numbers were 167. 600. 681. It is therefore apparant

(3)

rant, that the fourth proportionall, which I seeke for, muſt likewiſe increaſe aboue the totall *Sine;* ſo that his Logarithme is leſſe then nothing, and the number anſwering to it is not to be found in this Table. Therefore by the 4. *chap.* 1. *lib.* the 9 *Sect.* *pag* 18. I doe adde 23025 (cutting off here alſo the two laſt figures) to the Logarithme of 680 : therefore the third number being 68, and not 680. his Logarithme by this new increaſe is 28134. And becauſe by the 5 *prop.* 2. *chap. lib* .1. the Logarithmes of the two middle numbers are equall to the Logarithmes of the two extreame proportionals, therefore out of 32077, the ſumme of the two middle Logarithmes I take 17928, the Logarithme of the firſt, and there remaineth 14149. the Logarithme of the fourth proportionall, which I deſire : to the which, in the Table, 243 doth anſwere. Therefore I ſay, as before, that the proportionall part to bee added to the arke found, is $24\frac{3}{10}$. For though the fourth proportionall 243 be of 3 figures, yet wee may preſently diſcerne that the cauſe of this greatneſſe in this laſt proportionall is becauſe the one of the two middle numbers is too great, which ſhould be onely 60 and 68, and that if here we cut off the laſt figure, then will the foure numbers keepe due proportion 167. 60. 68. $24\frac{3}{10}$.

3. The third way of finding this proportionall part, is by this triangular table which was drawne by M^r. *E. Wright* ; and becauſe at his death he left no deſcription of it, nor manner how to vſe it, at the requeſt of ſome friends, I make bold to ſupply his place ſo well and plainly as I can.

You

(4)

You fee then in this triangle, three forts of
lines, fome paralell to the bafe AB, others
perpendicular vnto it, and the third fort
drawne all from the verticall angle C, vnto
equall partes of the bafe. Thefe laft may
be called *Diagonall* lines.

You fee likewife the bafe AB diuided
into 60 equall parts, and a line vnder it, and
paralell vnto it into 1000.

In like fort, the perpendicular line CB
is diuided into 1000, and vpon the inter-
fections of the paralells, and the Diagonall
CA are fet the fame numbers that are fet vpon the other ends of the paralels, in the perpendicular CB. Amongft thefe numbers we
muft place our three giuen numbers, and by
helpe of the lines we fhall amongft the fame,
finde out Geometrically, the fourth proportionall, which we defire.

For Example.

Take the fame numbers wee had before
167, and 60. the two Tabular Differences,
and 68 the Occurring difference. Of thefe
there are two which are differences of Logarithmes, to wit, 167, and 68, the firft and
the third : thefe being of one kinde, or Homogeneall, are to haue like fituation in the
Triangle. And the fecond being Homogeneall to the fourth, which is fought for, is
to bee placed on a differing fide from the o-
ther two, vpon which differing fide the fourth
proportionall is to bee found. As here I take
167, and 68, and count them from the poynt
C in the Diagonall line CA, and fuppofing
a perpendicular line to bee drawne from the
end of the leffe number till it cut the pa-
ralel

(5)

ralell line drawne from the end of the greater number, by this poynt of Interſection
I drawe an imaginarie Diagonall line from
the poynt C, till it cut the Baſe AB, and
counting from B to that Diagonall, I finde
24, and about $\frac{1}{2}$, which is the part proportionall I deſire, as in the former operations.

But becauſe 167 and 68 are ſuch ſmall numbers, and fall ſo neere the angle, therefore
the concourſe of the paralell and perpendicular is not ſo exactly diſcerned, and the
whole operation is more troubleſome and vncertaine. It is therefore conuenient in ſuch
caſes, to take the double, or treble of both
theſe giuen numbers, or the halfes, or any like
parts of them both : and to enter the
Table with theſe other numbers, in ſtead
of the former, proceeding in all things as before; then ſhall we, when the numbers reach
neerer vnto the Baſe, finde the poynt of
concourſe, and the part proportionall more
exactly and eaſily then before. As if we take
835 and 340 the quintuples of the firſt and
third numbers, wee ſhall more plainely
and diſtinctly finde the fourth proportionall
to be 24$\frac{1}{3}$.

In like ſort, page 52, line 36, I would finde
the differentiall anſwering to 16° : 24 : 27.
and becauſe the Table extendeth but to minuts, I muſt finde the part proportionall anſwering to 27. Here the firſt giuen number is
the Tabular differēce of minuts 60: the third
number is the Occurring Homogeneall difference 27. The ſecond giuen number is the
Tabular difference of the Differentials
1074. Theſe three I place thus : The firſt 60.

K   and

(6)

and the third 27, being *Homogeneall* vnto it, I
place vpon the bafe AB : and becaufe the fecond 1074 is too great for the table, & if one
figure be cut away 107 remaining wil fall vpon the Diagonall line CA (on the which it is
to be placed) too neere to the angle C, therefore I take 537, the halfe of the fecond, and
place it vpon the Diagonall line CA, drawing
a paralell from that poynt, till it meete with
the Diagonall comming from 27, and from the
concourfe of that paralell with this Diagonal,
I draw a perpendicular vpwards, till it cut
the Diagonall line of 66. or the line CA,
and I finde that the diftance of this interfe-
ction from C, is about 240, which is the halfe of
the fourth proportionall, becaufe 537 is the
halfe of the fecond giuen number *Homogeneall* vnto this fourth. I fay therefore that the
part proportionall anfwering vnto 27, is 480.
which being fubducted from 1223101, the differentiall of 16° : 24, there remaineth 1222621
for the differentiall 16° : 24, 27, which differeth fomewhat from that differentiall which is
fet downe in the booke : the reafon whereof is,
becaufe in fo fmall a Table it is impoffible to
difcerne all the feuerall parts, the omiffion of
which will make no fenfible difference in any
worke. If in feeking the fquare roote, or cubicke roote of 19, I would find the Logarithme
of 190000, feeking this number amongft the
Sines, I cannot finde it, but at 10° : 57 I finde
189952, which is leffe then the giuen number
by 48. This is the *Occurring difference.* The tabular difference *Homogeneall* to this, is 286.
The other tabular difference of the Logarithmes is 1502 : thefe differences are giuen,
which being placed in due order, the firft of
them

(7)

them is 286. the fecond 1502, the third 48, and
that the firft & third number may fall neerer
to the bafe, I double them both, fo haue I 572,
and 96, thefe I place, in ftead of the other giuen numbers, vpon the perpendicular CB, or
on the Diagonall CA. Likewife becaufe the
fecond number 1502 is too great, I take the
the halfe of it, fo haue I 751, which I place vnder the bafe AB, vpon the line diuided into
1000 : and from that point draw a diagonall,
till it meete with the paralell of 96. the third
number, and from the poynt of concourfe
with that paralell, I draw a perpendicular, till
it croffe the paralell of 571 the firft number.
By this laft concourfe of the perpendicular
and paralell, I draw an other diagonall, cutting
the line vnder AB, in which the fecond number was counted, and the parts 125 betwixt
that poynt and the end of the line towards B,
being doubled, (becaufe the fecond being Homogeneall to this, was halfed) are the fourth
proportionall required, which may bee 250.
and this proportionall being taken away from
the tabular Logarithme 1660982, leaueth
1660732 for the Logarithme of 190000.
    The fame may be performed, if the firft and
third be placed on the bafe; and the fecond
vpon the fide line, thus; Draw two Diagonals
from thofe two points of the bafe, & from the
concourfe of the Diagonall of the third with
the paralell of the fecond, draw a perpendicular vpward, till it meete with the Diagonall
of the firft; the paralell paffing by that point,
fhall in the line CB fhew the fourth proportionall.
    Thus may wee inftrumentally come fome-
what neere to that which wee defire, efpeci-
cially

(8)

cially if vpon a faire large Paſte-boord
wee make a great Triangle, curi-
ouſly diuided and lined, according to this pat-
terne.

FINIS.

Errata in the Treatiſe.
Pag. 15. lin. 28. make it Tangēts 1370505
    lin. 29. Make it number 1370305
Pag. 19. lin. 25. reade   4605168—00
Pag. 21. lin. 33. reade   ——34914—00
Pag. 29. lin. 10. reade   ——693147
Pag. 45 .lin. 4. reade ZSP
Pag. 72. lin. 30. reade ZP.
Pag. 75 lin, 17. reade halfe the aggregate.